Electrician's Guide to Control and Monitoring Systems

Installation, Troubleshooting, and Maintenance

Albert F. Cutter, Sr.

New York   Chicago   San Francisco   Lisbon   London   Madrid
Mexico City   Milan   New Delhi   San Juan   Seoul
Singapore   Sydney   Toronto

The McGraw·Hill Companies

Cataloging-in-Publication Data is on file with the Library of Congress

Copyright © 2010 by Albert F. Cutter, Sr. All rights reserved. Printed in the United States of America. Except as permitted under the United States Copyright Act of 1976, no part of this publication may be reproduced or distributed in any form or by any means, or stored in a data base or retrieval system, without the prior written permission of the publisher.

1 2 3 4 5 6 7 8 9 0   DOC/DOC   1 6 5 4 3 2 1 0

ISBN 978-0-07-170061-0
MHID 0-07-170061-7

Sponsoring Editor
Joy Bramble

Editorial Supervisor
Stephen M. Smith

Production Supervisor
Richard C. Ruzycka

Acquisitions Coordinator
Michael Mulcahy

Project Manager
Vipra Fauzdar, Glyph International

Copy Editor
Patti Scott

Proofreader
Laura Bowman

Indexer
Robert Swanson

Art Director, Cover
Jeff Weeks

Composition
Glyph International

Printed and bound by RR Donnelley.

McGraw-Hill books are available at special quantity discounts to use as premiums and sales promotions, or for use in corporate training programs. To contact a representative, please e-mail us at bulksales@mcgraw-hill.com.

This book is printed on acid-free paper.

Information contained in this work has been obtained by The McGraw-Hill Companies, Inc. ("McGraw-Hill") from sources believed to be reliable. However, neither McGraw-Hill nor its authors guarantee the accuracy or completeness of any information published herein, and neither McGraw-Hill nor its authors shall be responsible for any errors, omissions, or damages arising out of use of this information. This work is published with the understanding that McGraw-Hill and its authors are supplying information but are not attempting to render engineering or other professional services. If such services are required, the assistance of an appropriate professional should be sought.

This book is dedicated to my friends, brother and sister electricians who have encouraged me to put my knowledge onto paper, and to my late wife, Xiang Yun, whose love gave me the strength to believe in myself.

ABOUT THE AUTHOR

Albert F. Cutter, Sr., is owner and CEO of Intuitive Technologies, Inc., and GMP of AppsDev LLC. A union electrician—Local 456 of the International Brotherhood of Electrical Workers (IBEW)—for more than 40 years, he has extensive experience working on control systems and automation. Mr. Cutter has taught in the apprenticeship program of the IBEW and has also taught college-level courses on production control.

Contents

Foreword vii
Preface ix
Acknowledgments xi

Chapter 1. Ladder Diagrams — 1

Emergency Power Off (EPO) System Ladder Diagram — 3
EPO Procedure — 19
Punch Press — 20
Punch Press Procedure — 28

Chapter 2. Input Devices — 31

Push Buttons and Selector Switches — 32
Automatic Float Switch — 60
Pressure Switch — 62
Temperature Switch — 65
Photoelectric Sensor — 68
Inductive Proximity Sensors — 71
Capacitive Proximity Sensors — 74
Limit Switches — 77

Chapter 3. Output Devices — 81

Relays — 82
Open Frame Relay — 83
Heavy-Duty Industrial Relays — 85
Sealed Relay (Ice Cube) — 87
Latching Relays — 90
Timing Relays — 92
Solid-State Relays — 94
Solid-State Timer Relays — 97
Shunt Trip Breaker — 99
Three-Phase Motor Starter — 100

Chapter 4. Monitoring Systems — 103

    Programmable Logic Controllers — 104
    Monitoring and Control Modules — 105
    RS-232 Introduction and Specifications — 106
    RS-485 Introduction and Specifications — 114
    Modbus Introduction and Specifications — 122
    Kilowatt Meter — 126
    Ethernet Introduction and Specifications — 127
    Universal Serial Bus — 129

Chapter 5. Terminology and Definitions — 131

    Appendix A Motor Control—3 Phase 143
    Appendix B Ladder Diagrams 145
    Appendix C DGH Corporation Modules 155
    Appendix D Electrical Control Symbols 167
    Index 189

Foreword

Over the course of my 25-year career in the electrical industry, I have seen the scope of work that defines the electrician grow significantly. Today, the knowledge, skills, and abilities required for electricians who install, troubleshoot, and maintain electrical systems are mind-boggling. In addition to the expanding scope of work, the pace at which change occurs and the pace at which technology improves are unfathomable.

Even a cursory review of today's electrical market will reveal that nowhere is the confluence of these two developments more apparent than in the area of electrical/electronic systems control and monitoring. Albert Cutter has done an outstanding service for our industry by developing a hands-on reference and training tool in his new work, *Electrician's Guide to Control and Monitoring Systems*.

Electrical workers today are faced with the increased challenge of installing, troubleshooting, and maintaining the systems that control and monitor critical-mission processes and operations. The *Guide* provides detailed drawings and step-by-step explanations that will be extremely useful to field and plant electricians who are required to install and maintain these systems. Extensive coverage is provided on input and output devices, ladder diagrams, pilot devices, and electrical control and monitoring symbols.

Today's electrical industry is a dynamic one that relies heavily on the latest electrical/electronic systems to control and monitor processes that are critical to so very many industries, including the automotive, pharmaceutical, petrochemical, and engineering communities. Electricians must be trained and capable of meeting the increasing demands that these industries place on them. Mr. Cutter's *Guide* will go a long way toward helping the next generation of electricians meet this challenge.

Michael I. Callanan
Executive Director
National Joint Apprenticeship and Training Committee (NJATC)

Preface

This book contains practical information needed by the electrical contractor, electrician, or maintenance worker to understand and work with the technology used in systems found today. The book focuses on control and monitoring systems. The information provided here will assist in the installation, troubleshooting, and maintenance of these systems.

This information is presented with as little jargon as possible and with as many examples as are practical. There are hundreds of product photographs and information sheets and line drawings. Ladder diagrams are broken down and rebuilt to allow the reader to have a better understanding of the symbols and language used in them.

Network systems are explained in plain language. The only theory used is what is absolutely necessary to understand the subject. Sample devices and circuits are used to help the reader understand control and monitoring systems. All the relevant 2008 National Electrical Code® articles are explained.

Albert F. Cutter, Sr.

Acknowledgments

I want to thank the friends and brothers who have been an invaluable resource in writing this book: Jaime Lim, Pat Lyons, Miguel de Leon, William O'Sullivan, and Gary Menghi. Thank you for your help and support.

I also want to acknowledge

Rockwell Automation, Inc.
1201 South Second Street
Milwaukee, WI 53204-2496
www.ab.com

DGH Corporation
P.O. Box 5638
Manchester, NH 03108-5638
(603) 622-0452
www.dghcorp.com

Tom Henry, *Dictionary for the Electrician with Formulas,* copyright 1997

Wikipedia, the free encyclopedia
http://en.wikipedia.org/wiki/Main_Page

National Electrical Manufacturers Association (NEMA) Standards Publication ICS 19-2002

2008 National Electrical Code (NEC)

And special thanks go to the following people:

Stacey Iwinski, Intellectual Property, Rockwell Automation, Inc.

David Dutile, President; David Reed; and Lea Levesque, DGH Corporation

Michael G. McLaughlin, President
Joseph V. Egan, Business Manager
Francis T. Leake, Assistant Business Manager
International Brotherhood of Electrical Workers, Local 456
1295 Livingston Avenue
North Brunswick, NJ 08902

Chapter 1

Ladder Diagrams

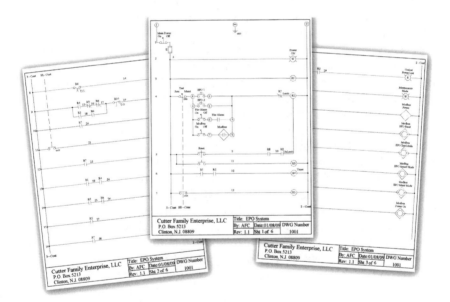

Ladder diagrams are used to present the electrical wiring in a control circuit. The diagram is made up of vertical lines that represent the control voltage and horizontal lines, or rungs, that represent the control circuits. The left line is the high-voltage or plus leg, and the right is the neutral or minus leg of the circuit. The control voltage can be AC or DC and can be any value needed but is usually from 0.5 to 220 V AC.

A rung is read from left to right. The inputs, contacts, or control devices are shown on the left. It is important to note that there are an unlimited number of input devices. Some are mechanical, others are electrical, and still others are solid-state. The one thing that you should understand is that whatever it is, it's just a switch. It is either Open or Closed. There may be some external action that controls the input; e.g., a time-closed contact will close at a preset time after the timer is energized. But when it has completed its cycle, it is closed—no magic. The outputs are shown on the right of the rung. The exception—there is always an exception—is the motor controller overload contacts. They are always shown Normally Closed and are the last element(s) in the rung.

It is not possible and would be confusing to use pictures of the devices on the diagram. So standard symbols were developed. Symbols are the language of ladder diagrams; they depict the inputs and outputs of the circuit. The inputs are always shown in the off or no-voltage state unless otherwise noted on the diagram. However, they can be shown with mechanical devices such as springs holding them opened or closed. For example, a push button is shown in the neutral state, not pushed. A limit switch can be shown in a held Open or Closed state; its symbol will indicate this. The National Electrical Manufacturers Association (NEMA) Standards Publication ICS 19-2002 sets forth the standards for the use of symbols. A copy can be obtained at www.nema.org.

Not all engineers and manufacturers use standard symbols. So look for notes on the drawing for an explanation of unknown symbols. If you familiarize yourself with the electrical control symbols shown in detail in App. D and on this book's website, you will be able to read most diagrams.

Figure 1-1 shows a simple ladder diagram. First we assume that there is voltage between the line L1 and the neutral N. From left to right the push button is Normally Open so the pilot light is off. When the push button is pressed, contact closes and the green pilot light turns on.

Figure 1-1

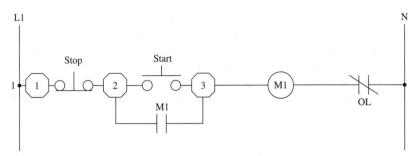

Figure 1-2

Figure 1-2 is the classic three-wire motor control circuit. It is called a three-wire circuit because the circuit has three wires that are external to the motor control box.

At the left of the circuit is the number 1. This is the line number. I like to use line numbers for the primary rungs so that it is easier to communicate on the telephone or via e-mail. Next we have L1, which is the feed wire. Then we have an octagon symbol with the number 1 in it. This is the symbol I use for a terminal on a terminal strip. Different designers might use a rectangle, square, or circle. It will be obvious when you look at the diagram, and it should also be stated in the notes section of the diagram. The Normally Closed push button is used for the Stop button; when pressed, it opens the circuit. Next we have terminal 2. To the right is the Normally Open push button that is the Start button. When pressed, it completes the circuit and energizes the motor contactor coil M1. On the next line down is the Normally Open contact of motor contactor coil M1, as indicated by the M1 next to the contact. A coil or relay can have many contacts. They can be Normally Open or Closed; the symbol will indicate this. Then there is terminal 3 that connects to the Start button and the coil. The motor contactor coil M1 is the coil of the motor starter; when energized, it will close the contacts and start the motor. The last element in the rung is always the overload contact of the motor control starter. The circuit will be opened by the overload contact if the motor circuit rises above a preset current. Let's recap. Push the Start button; it will close the circuit to the coil M1, and this will complete the circuit. This will close the contact M1, and the circuit will remain energized until the Stop button is pressed or the control power is lost. The electrically latched relay circuit will fail-safe so the motor will not come on when the power is restored.

Emergency Power Off (EPO) System Ladder Diagram

We will take a look at a more complex control circuit, using the diagram for an EPO system to illustrate the use of ladder diagrams and how to read them. NEC Section 645.10 requires that a computer room must have an EPO system to be classified as an IT Equipment Room under NEC 645. This system must disconnect all electrical power in the computer room. This is typically done by using shunt trip breakers on all the power circuit feeders. The system will have mushroom push buttons by each exit from the computer room. It may also be connected to a remote system or smoke detector panel to trigger the EPO system. Panel documentation

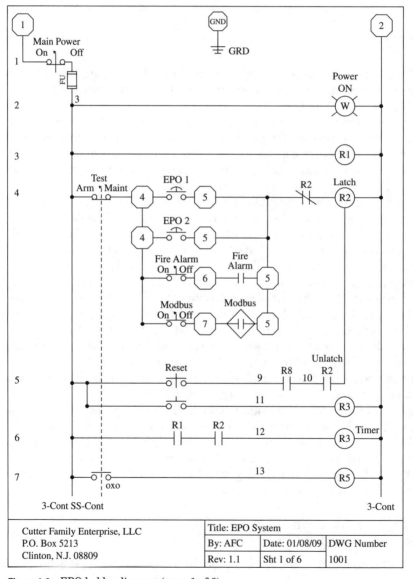

Figure 1-3 EPO ladder diagram (page 1 of 6).

Ladder Diagrams 5

with this system comprises six pages. Pages 1, 2, and 3 are ladder diagrams (Figs. 1-3 to 1-5); page 4 is a contact layout and notes; page 5 is a bill of materials; and page 6 is the panel cover layout. For the complete package please see App. B.

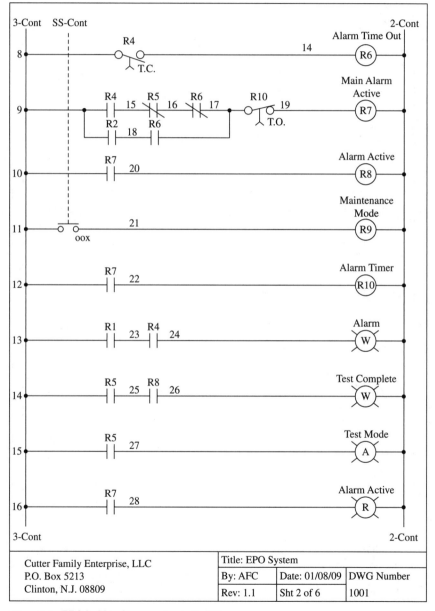

Figure 1-4 EPO ladder diagram (page 2 of 6).

6 Chapter One

There are three basic modes of operation:

Armed mode. System is energized, and the system is standing by for an alarm or reset.

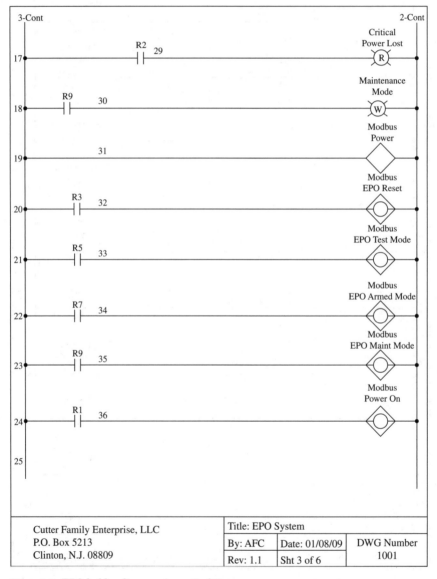

Figure 1-5 EPO ladder diagram (page 3 of 6).

Test mode. System is in the armed state, but the output contacts have been disconnected and will not close on alarm. This can be used to test the functions of the system without turning off the power or equipment.

Maint mode. System will not function, all controls will be in the neutral state, and the alarm will not function.

Figure 1-6

Shown in Fig. 1-6:

Line 1 On the left we have terminal 1, which is the hot leg, or L1, of the 120-V AC control circuit. On line 1 on the left is a key switch to turn the power on or off to the system.

The next element is the fuse. NEC 2008 Article 430-72 requires that there be overcurrent protection for the control circuit. The control wiring is typically 14 gauge wire rated at 15 A (NEC 2008 Article 240.3 D-3) and is protected typically by a 10-A fuse.

The next element is the ground connection. The enclosure and the back plate of the panel must be grounded. Next we have terminal 2, which is the neutral leg, or L2, of the control circuit.

Figure 1-7

Shown in Fig. 1-7:

Line 2 The Power On indicator is a white pilot light that will be on whenever the system control power is on or energized.

8   Chapter One

Figure 1-8

Shown in Fig. 1-8:

Line 3   R1 is an 8-pole relay that can have any contact changed in the field from Normally Open to Normally Closed. See the bill of materials in Fig. 1-34. The relay will be on whenever the system control power is on or energized. This relay will be used in the circuit, but its primary use is to monitor the control voltage. One Normally Open contact is on line 6 and one is on line 24. One Normally Open contact is for external system monitoring of the AC power terminals 40 and 41. Four Normally Open contacts are for future-use terminals 42 to 49.

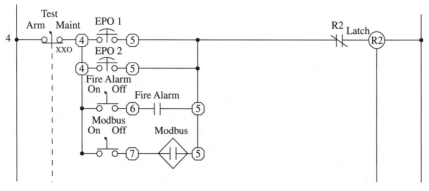

Figure 1-9

Shown in Fig. 1-9:

Line 4   The first element is the three-position key Selector switch. The three positions are arm, test, and maint. The XXO below the contact shows that the contact is closed in the arm and test positions. Next is terminal 4, then the first of the Normally Open Mushroom keyed release push buttons. Any number of contacts can be used to alarm the system; just put them in parallel with the current contacts.

The next line down is another Mushroom push button. NEC 645 requires that there be an alarm at each exit.

The next line down is the keyed Selector switch for the fire alarm interface contact. If there is a test of the fire alarm system, the panel can be set by the key switch to bypass the alarm. Next we have terminal 6. Then the next element is a contact from the fire alarm system that will close when the fire alarm is active.

The next element is terminal 5.

The next line down is the interface for the RS-485 module with Modbus communication. See the Modbus section of Chap. 4. The Modbus module interfaces the EPO system with a control center. If the keyed Selector switch is closed, the Modbus module will give the system monitoring personnel or software to alarm the panel if necessary. The next element we have is a Normally Open output from the Modbus.

The next element in line 4 is a Normally Closed contact from R2. This will open when relay R2 is latched. The next element is the R2 latching relay. The latching relay is mechanically held in place even if the power fails. The only way to reset the relay is to apply power to the unlatching coil. R2 has Normally Open contacts in lines 5, 6, and 17 and a Normally Closed contact in line 4.

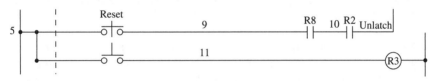

Figure 1-10

Shown in Fig. 1-10:

Line 5 The first element is the Normally Open contact from the reset push button. When pushed after an alarm, the push button will reset and unlatch the R2 relay. The next element is a Normally Open contact from relay R8 on line 10. The next element is a Normally Open contact of R2, and it will be closed if the relay is latched.

In the next line down, the first element is a Normally Open contact of the Reset push button. Closing the push button will energize R3 which is the next element. R3 has a Normally Open contact on line 20.

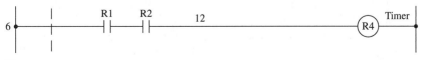

Figure 1-11

Shown in Fig. 1-11:

Line 6 The first element is a Normally Open contact of R1 which will be closed if the panel has power. The next element is a Normally Open contact of R2. The next element is time delay relay R4. The logic is that if the panel has power and the R2 relay is latched, then the timer will start. R4 has a Normally Open contact on lines 9 and 13. The Normally Open, time-closed contact is on line 8.

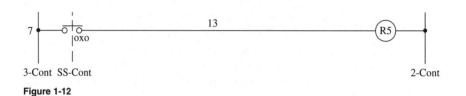

Figure 1-12

Shown in Fig. 1-12:

Line 7 The first element is the Selector switch. The OXO indicates that the contact is closed in the test mode. The next element is the R5 relay. The R5 relay has Normally Open contacts in lines 14, 15, and 21 and a Normally Closed contact in line 9. The logic is that when the Selector switch is in the center position, test mode, relay R5 will be energized. The 3-Cont, SS-Cont, and 2-Cont indicate that the number 3 and 2 wires and Selector switch mechanical operator continue to the next page.

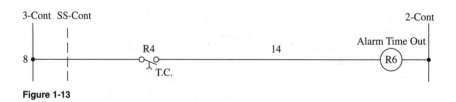

Figure 1-13

In Fig. 1-13, the 3-Cont, SS-Cont, and 2-Cont indicate that the number 3 and 2 wires and the Selector switch mechanical operator continue from the previous page.

Line 8 The first element is a Normally Open, time-closed contact from relay R4. The next element is relay R6. The logic is that after relay R4 has timed out, the contact will close, and relay R6 will be energized. Relay R6 has a Normally Open contact on line 9 and a Normally Closed contact on line 9.

Ladder Diagrams 11

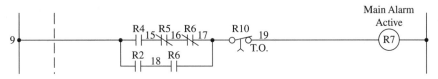

Figure 1-14

Shown in Fig. 1-14:

Line 9 The first element is a Normally Open contact of relay R4. The next element is a Normally Closed contact of relay R5. The next element is a Normally Closed contact of relay R6. In the next line down, the first element is a Normally Open contact of relay R2. The next element is a Normally Open contact of relay R6. The next element is a Normally Closed time-open contact of time delay relay R10. The logic is that if relay R4 is closed and relay R5 is closed and relay R6 is closed or relay contact R2 is closed and relay contact R6 is closed and relay contact R10 is closed, then relay R7 will be energized. Relay R7 is the main alarm relay and triggers the output contacts. Relay R7 has Normally Open contacts on lines 10, 12, 16, and 22. Three Normally Open output contacts are on terminals 48 to 53. These contacts are for the CPC shunt trip circuits. One Normally Open contact is on terminals 54 and 55, a fire alarm system monitoring circuit.

Figure 1-15

Shown in Fig. 1-15:

Line 10 The first element is a Normally Open contact of relay R7, and the next element is relay R8. The logic is that when R7 closes, relay R8 will be energized. This relay is needed to give more contacts 8 to the function of the R7 relay. Relay R8 has Normally Open contacts on line 26. Three Normally Open output contacts are on terminals 58 to 63. These contacts are for the CRAC shunt trip circuits. One Normally Open contact is on terminals 64 and 65, a fire alarm system damper shutdown.

12   Chapter One

Figure 1-16

Shown in Fig. 1-16:

Line 11 The first element is a Normally Open contact of the Selector switch. The OOX indicates Selector switch maint (maintenance) mode. The next element is relay R9. The logic is thus: When the Selector switch is in the maint mode position, then relay R9 will be energized. Relay R9 has a Normally Open contact on lines 18 and 23 and a Normally Open contact on terminals 66 and 67. The Selector switch is in the maint position, all controls are inactive, and the pilot light will be energized. Also, the Modbus module will be altered.

Figure 1-17

Shown in Fig. 1-17:

Line 12 The first element is a Normally Open contact of relay R7. The next element is the time delay relay R10. The logic is thus: When the panel is in alarm, relay R7 will be energized and start the timer R10. Relay R10 has a time-open contact in line 9.

Figure 1-18

Shown in Fig. 1-18:

Line 13 The first element is a Normally Open contact of relay R1. The next element is a Normally Open contact of relay R4. The next element is a white pilot light. The logic is that when the panel has control voltage and the time delay relay R4 closes, the alarm pilot light will be illuminated.

Ladder Diagrams   13

Figure 1-19

Shown in Fig. 1-19:

Line 14 The first element is a Normally Open contact of relay R5. The next element is a Normally Open contact of relay R8. The next element is a white pilot light. The logic is that when relay R5 is energized in test mode, and relay R8 is energized, the Test Complete pilot light will be illuminated.

Figure 1-20

Shown in Fig. 1-20:

Line 15 The first element is a Normally Open contact of relay R5. The next element is an amber pilot light. The logic is that when the relay R5 is energized in test mode, the Test Mode pilot light will be illuminated.

Figure 1-21

Shown in Fig. 1-21:

Line 16 The first element is a Normally Open contact of relay R7. The next element is a red pilot light. The logic is that when the relay R7 is energized in alarm mode, the Alarm Active pilot light will be illuminated until the time delay of relay R10.

The 3-Cont and 2-Cont indicate that the number 3 and 2 wires continue to the next page.

Figure 1-22

In Fig. 1-22, the 3-Cont and 2-Cont indicate that the number 3 and 2 wires continue from the previous page.

Line 17 The first element is a Normally Open contact of latching relay R2. The next element is a red pilot light. The logic is that when the latching relay R2 is energized in alarm mode, the Critical Power Lost pilot light will be illuminated.

Figure 1-23

Shown in Fig. 1-23:

Line 18 The first element is a Normally Open contact of relay R9. The next element is a white pilot light. The logic is that when the Selector switch is in the maint mode position, the relay R9 is energized. The Maintenance Mode pilot light will be illuminated.

Figure 1-24

Shown in Fig. 1-24:

Line 19 Power for the Modbus unit. The Modbus unit allows an interface to the control and monitoring system.

Ladder Diagrams   15

Figure 1-25

Shown in Fig. 1-25:

Line 20   The first element is a Normally Open contact of relay R3. The next element is the Modbus EPO Reset input. The logic is that when the system has been reset using the reset push button, the Modbus will receive the indication that this event has accorded and can be reported to the monitoring system.

Figure 1-26

Shown in Fig. 1-26:

Line 21   The first element is a Normally Open contact of relay R5. The next element is the Modbus EPO test input. The logic is that when the system has been put in the test mode using the Selector switch, the Modbus will receive the indication that this event has accorded and can be reported to the monitoring system.

Figure 1-27

Shown in Fig. 1-27:

Line 22   The first element is a Normally Open contact of relay R7. The next element is the Modbus EPO armed input. The logic is that when the system has been put in the armed mode using the Selector switch and the system is alarmed using any of the devices, the Modbus will receive the indication that this event has accorded and can be reported to the monitoring system.

Figure 1-28

Shown in Fig. 1-28:

Line 23 The first element is a Normally Open contact of relay R10. The next element is the Modbus EPO maint mode input. The logic is that when the system has been put in the maint mode using the Selector switch, the Modbus will receive the indication that this event has accorded and can be reported to the monitoring system.

Figure 1-29

Shown in Fig. 1-29:

Line 24 The first element is a Normally Open contact of relay R1. The next element is the Modbus EPO power on input. The logic is that when the system has been put in control power, the Modbus will receive the indication that this event has accorded and can be reported to the monitoring system.

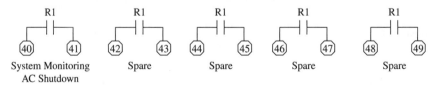

Figure 1-30

Figure 1-30 shows that R1 will close when there is power to the system panel circuit.

Ladder Diagrams 17

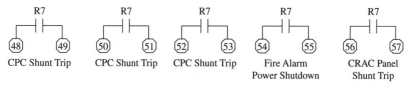

Figure 1-31

Figure 1-31 shows that R7 and R8 will close when there is an alarm. This will energize the shunt trip breakers and turn off all power to the CRAC units.

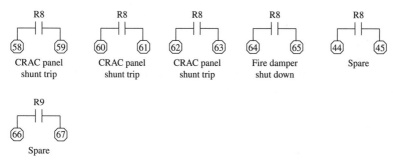

Figure 1-32

Figure 1-32 shows that R8 will close when there is an alarm, and R9 will close when the system is in the maint mode.

> **Notes:**
>
> 1. ① Indicates a terminal on the TB1 terminal strip.
>
> 2. After an alarm the restoring panel will start a timer R4. You will have 5 to 10 min to reset the panel or the power will turn off and recycle the panel. Recommended time is 5 min.
>
> 3. Timer R10 will open the circuits to the shunt trip breakers after x (time recommended 10 s). This will remove the power from the controlled devices, rendering them safe.

Figure 1-33

A notes section such as the one shown in Fig. 1-33 will have information that you cannot perceive by looking at the ladder diagram.

Bill of Materials			
Item	Qty.	Manufacturer	Part Number
4-Pole Relay	8	Allen-Bradley	700-N400A
4-Pole Time Delay Relay	2	Allen-Bradley	700-NT400A
4-Pole Contact Front Deck	10	Allen-Bradley	700-NA00
4-Position Relay Rack	2	Allen-Bradley	700-NP4
Terminal Block	55	Allen-Bradley	1492-CA2
Terminal Rail	1	Allen-Bradley	1492-CA2175(88/Rail)
Fuse Block	1	Allen-Bradley	1492-H6
2-Position Keyed Selector Switch	2	Allen-Bradley	800T-H33D1
3-Position Keyed Selector Switch	1	Allen-Bradley	800T-J44A
EPO Mushroom EPO Push Button	2	Allen-Bradley	800T-E1 5M6A
2-Pole Push Button Black Reset	1	Allen-Bradley	800T-A2A
Pilot Light, White	3	Allen-Bradley	800T-QH10W
Pilot Light, Red	2	Allen-Bradley	800T-QH10R
Pilot Light, Amber	1	Allen-Bradley	800T-QH10A
Wireway		Panduit	G1LG6
Wireway Cover		Panduit	C1lG6

Figure 1-34

A *bill of materials* is useful if you need to replace a device or get the specifications for a device (Fig. 1-34).

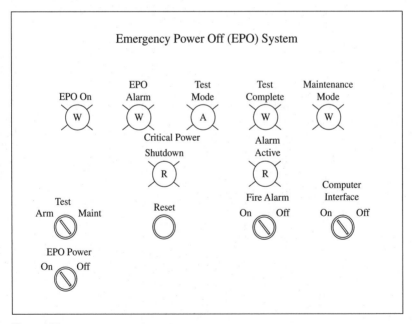

Figure 1-35

To locate the devices on the panel, refer to the *front panel display* (Fig. 1-35).

EPO Procedure

1. Use the keyed Selector switch to energize the EPO system; the Power On pilot light will be illuminated.
2. Use the keyed 3-position Selector switch to select any of the following modes:
    2.1. Armed mode
        2.1.1. The Armed pilot light will be illuminated.
        2.1.2. The Modbus module will be notified that the system is an armed event.
        2.1.3. The Mushroom buttons will be active.
        2.1.4. If the keyed Selector switch fire alarm is in the On position, the fire alarm system can activate the system alarm.
        2.1.5. If the keyed Selector switch computer interface is in the On position, the monitoring/control system can activate the system alarm using the Modbus module.
        2.1.6. When one or more of the above triggers sets the alarm:
            2.1.6.1. The output contacts for R7 and R8 will trip the shunt trip breakers and disconnect the power in the IT Equipment Room.
            2.1.6.2. The Alarm Active pilot light will be illuminated until the timer R10 times out after 10 s, cutting off power to the shunt trip breakers.
            2.1.6.3. The Critical Power Lost pilot light will be illuminated.
            2.1.6.4. The Modbus module will be notified of the alarm event.
            2.1.6.5. The power to the panels will be cut off by the shunt trip breakers.
            2.1.6.6. The EPO panel will be off.
        2.1.7. When power is restored:
            2.1.7.1. The system will alarm if the Reset button is not pushed within the time delay period of the R4 timer: 5 to 10 min.
            2.1.7.2. The Reset button is pushed.
                2.1.7.2.1. Relay R1 will reset/unlatch, and the system will return to Armed mode.
                2.1.7.2.2. The Armed pilot light will be illuminated.
                2.1.7.2.3. The Modbus module will be notified of the reset event.
    2.2. Test mode
        2.2.1. The Test mode pilot light will be illuminated.
        2.2.2. The R7 and R8 relays will be locked out, and the system output contacts will not close.

2.2.3. The Modbus module will be notified that the panel is test mode event.
2.2.4. The Mushroom buttons will be active.
2.2.5. If the keyed Selector switch fire alarm is in the On position, the fire alarm system can activate the system alarm.
2.2.6. If the keyed Selector switch computer interface is in the On position, the monitoring/control system can activate the system alarm using the Modbus module.
2.2.7. When one or more of the above triggers sets the alarm:
2.2.7.1. The output contacts for R7 and R8 will not trip the shunt trip breakers and disconnect the power in the IT Equipment Room.
2.2.7.2. The Alarm Active pilot light will be illuminated until the timer R10 times out after 10 s, cutting off power to the shunt trip breakers.
2.2.7.3. The Critical Power Lost pilot light will be illuminated.
2.2.7.4. The Modbus module will be notified of the alarm event.
2.2.7.5. The Test Complete pilot light will be illuminated.
2.2.8. When the test is complete:
2.2.8.1. The panel will alarm again if the Reset button is not pushed within the time delay period of the R4 timer.
2.2.8.2. The panel Reset is pushed.
2.2.8.2.1. Relay R1 will reset/unlatch, and the system will return to Armed mode.
2.2.8.2.2. The Armed pilot light will be illuminated.
2.2.8.2.3. The Modbus module will be notified of the Reset event.
2.3. Maint mode
2.3.1. In Maint mode:
2.3.1.1. The system will be inactive, and all controls will be disabled.
2.3.1.2. The Maintenance Mode pilot light will be illuminated.
2.3.1.3. The Modbus module will be notified of the alarm event.

Punch Press

The following is a ladder diagram for a hydraulic punch press. OSHA requires that the punch press have guards on all the openings to protect the operator when the system is running.

Ladder Diagrams 21

The press must shut down if the guards are opened.
There are three modes of operation:

Hand. In Hand mode the press can be inched or jogged, which will be used in the setup of the press production run.

Off. In Off mode the press will not operate.

Auto. The Auto mode will allow the press to operate in a continuous mode.

Figures 1-36 and 1-37 are sheets 1 and 2 of a three-sheet punch press ladder diagram.

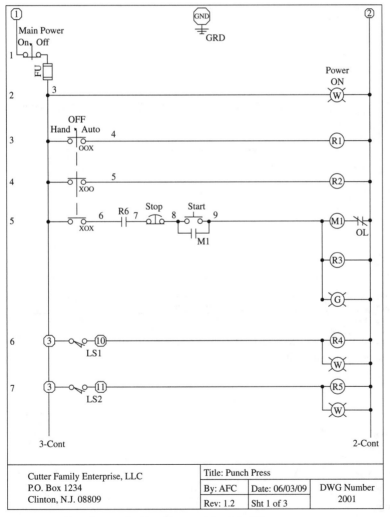

Figure 1-36 Punch press ladder diagram (sheet 1 of 3).

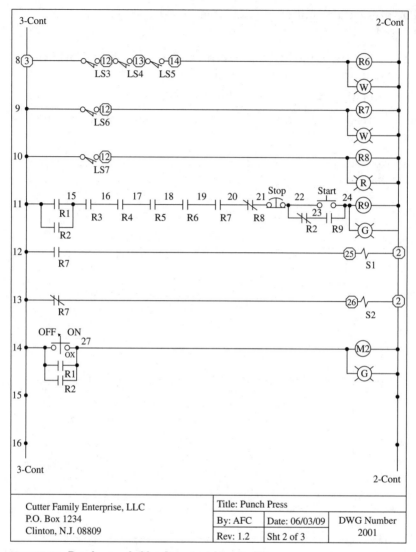

Figure 1-37 Punch press ladder diagram (sheet 2 of 3).

Figure 1-38

Ladder Diagrams   23

Shown in Fig. 1-38:

Line 1 On the left we have terminal 1, which is the hot leg, or L1, of the 120-V AC control circuit. On line 1 on the left is a key switch to turn the power on or off to the system.

The next element is the fuse. NEC 2008 Article 430-72 requires that there be overcurrent protection for control circuits. The control wiring is typically 14 gauge wire rated at 15 A (NEC 2008 Article 240.3 D-3) and is protected typically by a 10-A fuse.

The next element is the ground connection. The enclosure and the back plate of the panel must be grounded. Next we have terminal 2, which is the neutral leg, or L2, of the control circuit.

Figure 1-39

Shown in Fig. 1-39:

Line 2 The Power On indicator is a white pilot light that will be on whenever the system control power is on or energized.

Figure 1-40

Shown in Fig. 1-40:

Line 3 3-position Selector switch used to select Hand, Off, Auto. As indicated by the OOX, the Selector switch is closed in the Auto position and open in the Hand and Off positions. When closed, the relay R1 will be energized. This will close contacts in lines 11 and 14.

Figure 1-41

24 Chapter One

Shown in Fig. 1-41:

Line 4 3-position Selector switch used to select Hand, Off, Auto. As indicated by the XOO, the Selector switch is closed in the Hand position and open in the Auto and Off positions. When closed, the relay R2 will be energized. This will close contacts in lines 11 and 14.

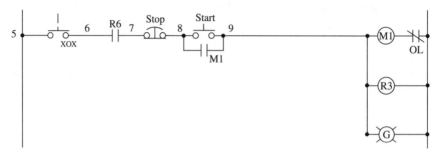

Figure 1-42

Shown in Fig. 1-42:

Line 5 3-position Selector switch used to select Hand, Off, Auto. As indicated by the XOX, the Selector switch is closed in the Hand and Auto positions and open in the Off position.

Next to the right is Normally Open contact R6. This contact will close when the motor, belt, and clutch covers are closed and will open and stop the motor if the covers are opened.

Next to the right is the Normally Closed Stop Red Mushroom push button. This button will be used to stop the motor once it is energized.

Next to the right is the Normally Open Green Start button. When closed, it will energize motor starter M1.

Next line down is the Normally Open contact of motor starter M1. When motor starter M1 is energized, this contact will close and latch.

Next to the right, motor starter M1 will be energized when the logic in line 5 is closed.

Next line down, relay R3 will be energized when motor starter M1 is energized, closing contacts in line 11.

Next line down, green pilot light G will be illuminated when motor starter M1 is energized.

Figure 1-43

Shown in Fig. 1-43:

Line 6 Terminal 3 on the terminal strip.
Next is limit switch LS1; this is closed when the front guard is closed.
Next is terminal 10 on the terminal strip.
Next to the right, relay R4 when energized will close contact in line 11.
Next line down, the white pilot light will be energized when relay R4 is energized.

Figure 1-44

Shown in Fig. 1-44:

Line 7 Terminal 3 on the terminal strip.
Next is limit switch LS2; this is closed when the rear guard is closed.
Next is terminal 11 on the terminal strip.
Next to the right, relay R5 when energized will close contact in line 11.
Next line down, white pilot light will be energized when relay R5 is energized.

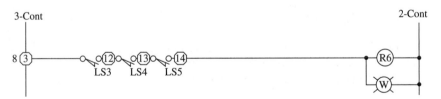

Figure 1-45

Shown in Fig. 1-45:

Line 8 Terminal 3 on the terminal strip.
Next is limit switch LS3; this limit switch is closed when the cover on the motor is closed.
Next is terminal 12 on the terminal strip.

26 Chapter One

Next is limit switch LS4; this limit switch is closed when the cover on the motor belt is closed.

Next is limit switch LS5; this limit switch is closed when the cover on the clutch is closed.

Next to the right, relay R6 when energized will close contact in line 11.

Next line down, white pilot light will be energized when relay R6 is energized.

Figure 1-46

Shown in Fig. 1-46:

Line 9 Limit switch LS6 is mounted on the material feeder. The limit switch will be closed when material is in the feeder.

Next to the right, relay R7 when energized will close contact in line 11.

Next line down, white pilot light will be energized when relay R7 is energized.

Figure 1-47

Shown in Fig. 1-47:

Line 10 Limit switch LS7 is mounted on the material output bin. The limit switch will be closed when material bin is full.

Next to the right, relay R8 when energized will close contact in line 11.

Next line down, white pilot light will be energized when relay R8 is energized.

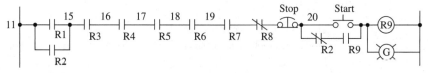

Figure 1-48

Shown in Fig. 1-48:

Line 11 Normally Open contact R1 will close when the 3-position switch is in the auto position.

Down one line, contact R2 will close when the 3-position switch is in the Hand position.

Up one line, Normally Open contact R3 will close when motor starter M1 is energized.

Next, Normally Open contact R4 will close when limit switch LS1 is closed when front guard is closed.

Next, Normally Open contact R5 will close when limit switch LS2 is closed when rear guard is closed.

Next, Normally Open contact R6 will close when limit switch LS3 is closed when motor cover is closed, LS4 when motor belt cover is closed, LS5 when the clutch cover is closed.

Next, Normally Open contact R7 will close when limit switch LS6 is closed when material feeder is closed.

Next, Normally Closed contact R8 will close when limit switch LS7 is closed when material output bin is full.

Next, Normally Closed Red Mushroom push button will unlatch relay R9.

Down one line, Normally Closed contact R2 will open when the 3-position select switch is in the Off or Hand positions. This will allow the press to be jogged or inched in the Hand mode.

Next, Normally Open contact R9 will close and latch the relay R9 when the Start push button is closed.

Up one line, Normally Open green Start push button will energize when the push button is pressed and the logic in line 11 is closed.

Next, relay R9 will energize when the logic in line 11 is closed or true.

One line down, green pilot light will illuminate when relay R9 is energized.

Figure 1-49

Shown in Fig. 1-49:

Line 12 Normally Open contact R9 will close when relay R9 line 11 is energized.

Next, terminal 25 on the terminal strip.

Next, solenoid S1 will be energized, which will close the clutch.

Figure 1-50

Shown in Fig. 1-50:

Line 13 Normally Closed contact R9 will open when relay R9 line 11 is energized.

Next, terminal 25 on the terminal strip.

Next, solenoid S2 will be deenergized, which will open the brake.

Figure 1-51

Shown in Fig. 1-51:

Line 14 2-position switch is closed in the On position and open in the Off position. When in the On position, this will energize motor starter M2 and run the press oiler.

Down one line, Normally Open contact R1 will close when the 3-position Selector switch is in the Auto position; this will energize motor starter M2 and run the press oiler.

Down one line, Normally Open contact R2 will close when the 3-position Selector switch is in the Hand position; this will energize motor starter M2 and run the press oiler.

Next, motor starter M2 when energized will run the press oiler.

Down one line, green pilot light will be illuminated when motor starter M2 is energized.

Punch Press Procedure

1. Use the keyed Selector switch to energize the punch press system; the Power On pilot light will be illuminated.
2. Use the keyed three-position Selector switch to select any of the following modes:
    2.1. Hand mode
        2.1.1. The motor can be started by pressing the Start button in line 5.
        2.1.2. Limit switch LS1 will energize relay R4 and white pilot light when front guard is closed.

- 2.1.3. Limit switch LS2 will energize relay R5 and white pilot light when rear guard is closed.
- 2.1.4. Limit switches LS3 motor cover, LS4 motor belt cover, and LS5 clutch cover are closed. This will energize relay R6 and white pilot light.
- 2.1.5. Limit switch LS6 will energize relay R7 and white pilot light when material is in the feeder.
- 2.1.6. Limit switch LS7 will energize relay R8 and white pilot light when material bin is full.
- 2.1.7. If all the logic in line 11 is closed, then the punch press will operate in the jog or inch mode.
- 2.1.8. Motor starter M2 will be energized and run the oiler.
- 2.2. Off mode
  - 2.2.1. The motor cannot be started.
  - 2.2.2. Limit switch LS1 will energize relay R4 and white pilot light when front guard is closed.
  - 2.2.3. Limit switch LS2 will energize relay R5 and white pilot light when rear guard is closed.
  - 2.2.4. Limit switches LS3 motor cover, LS4 motor belt cover, and LS5 clutch cover are closed. This will energize relay R6 and white pilot light.
  - 2.2.5. Limit switch LS6 will energize relay R7 and white pilot light when material is in the feeder.
  - 2.2.6. Limit switch LS7 will energize relay R8 and white pilot light when material bin is full.
  - 2.2.7. In line 11 Normally Open contacts R1 and R2 will be opened, and the punch press will not operate.
  - 2.2.8. Motor starter M2 will be energized and run the oiler if the 2-position Selector switch is in the On position.
- 2.3. Auto mode
  - 2.3.1. The motor can be started by pressing the Start button in line 5 if the logic in line is true.
  - 2.3.2. Limit switch LS1 will energize relay R4 and white pilot light when front guard is closed.
  - 2.3.3. Limit switch LS2 will energize relay R5 and white pilot light when rear guard is closed.
  - 2.3.4. Limit switches LS3 motor cover, LS4 motor belt cover, and LS5 clutch cover are closed. This will energize relay R6 and white pilot light.
  - 2.3.5. Limit switch LS6 will energize relay R7 and white pilot light when material is in the feeder.
  - 2.3.6. Limit switch LS7 will energize relay R8 and white pilot light when material bin is full.

30   Chapter One

2.3.7. If all the logic in line 11 is closed, then the punch press will operate in the Run mode.

2.3.8. Motor starter M2 will be energized and run the oiler.

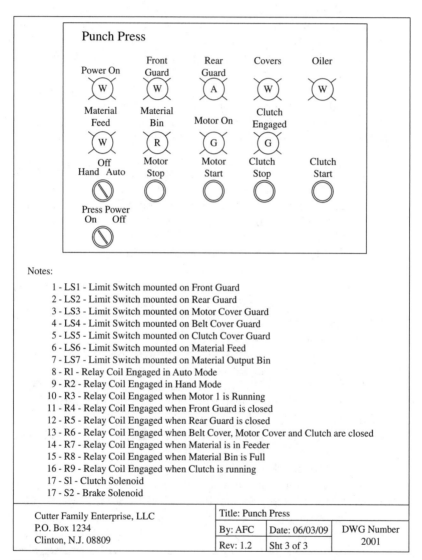

Figure 1-52

Shown in Fig. 1-52:

Front panel display and notes. This is useful to locate the devices on the panel and the functions of the system devices.

Chapter

2

Input Devices

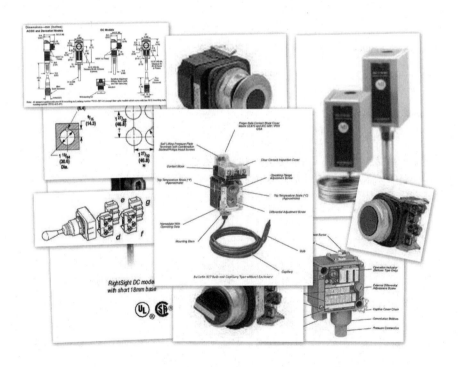

Chapter Two

I define an input device as a device that by some action controls the output of an electrical circuit. As you can imagine, there are an unlimited number of input devices. I will try to cover the most common devices. There are many manufacturers of these devices. I will for the most part use Rockwell Automation's Allen Bradley devices as I am most familiar with them and they are widely used in the field. The information contained here is not meant to replace the manufacturer's specifications for detailed data; please see the manufacturer's specifications.

Push Buttons and Selector Switches

The push button is a simple input device. The most basic is a single Normally Open contact. When the operator pushes the button, the circuit will close and complete the circuit.

Figure 2-1

In Fig. 2-1 the push button controls a pilot.

Figure 2-2 shows the breakdown of the 800T push button. The operator can be any of the operators described in this chapter.

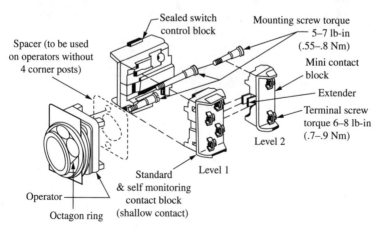

Figure 2-2 *Courtesy of Rockwell Automation, Inc.*

Input Devices

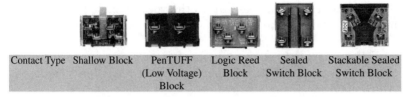

| Contact Type | Shallow Block | PenTUFF (Low Voltage) Block | Logic Reed Block | Sealed Switch Block | Stackable Sealed Switch Block |

Figure 2-3 *Courtesy of Rockwell Automation, Inc.*

Most manufacturers use a modular design that allows you to use the same contacts for different types of operators. Allen Bradley has the following contact blocks available for their operators:

Shallow block

Small block used where the mounting has limited depth or the door swing is an issue. (See Fig. 2-3.)

PenTUFF (low-voltage) block

Used for hazardous locations; types 7 and 9 explosion proof (type 3 and type 4 ratings available with accessories). (See Fig. 2-3.)

Logic reed block

Sealed reed switch can be used in hazardous locations. 800T operators using logic reed contact blocks and installed in a suitable enclosure are UL listed as suitable for use in Class I, Division 2/Zone 2 hazardous locations. (See Fig. 2-3.)

Sealed switch block

Sealed contacts used for hazardous locations. 800T operators using sealed switch contact blocks and installed in a suitable enclosure are UL listed as suitable for use in Class I, Division 2/Zone 2 hazardous locations. (See Fig. 2-3.)

Stackable sealed switch block

Sealed contacts used for hazardous locations. 800T operators using sealed switch contact blocks and installed in a suitable enclosure are UL listed as suitable for use in Class I, Division 2/Zone 2 hazardous locations. (See Fig. 2-3.)

These contacts can be stacked up to the recommended two deep, a total of four blocks with the operator.

Not all the operators shown on the following pages can use all the contact types. (For detailed specifications, see Allen Bradley at www.ab.com.)

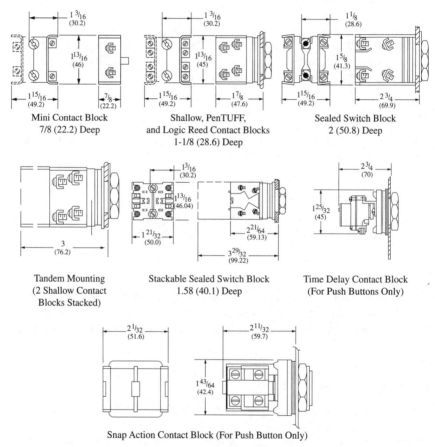

Figure 2-4 Dimensions in inches (millimeters). Dimensions are not intended to be used for manufacturing purposes. *Courtesy of Rockwell Automation, Inc.*

Dimensions for Allen Bradley 800T push buttons with contacts are shown in Fig. 2-4; these may change with different operators.

In Fig. 2-5 is shown the layout of the Allen Bradley 800T operators. This is the minimum distance between operators; your layout may be different.

Input Devices 35

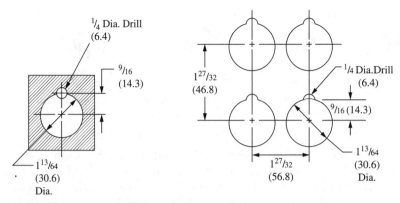

Figure 2-5 *Courtesy of Rockwell Automation, Inc. For detailed specifications, see Allen Bradley at www.ab.com.*

Note: A jumbo or large legend plate is recommended, if space allows.

| 2-Position Push-Pull | 2-Position Metal Push-Pull | 2-Position Push-Pull/Twist | 2-Position Push-Pull/Twist |
| Cat. No. 800T-FX6D4 | Cat. No. 800T-FXLE6D45 | Cat. No. 800T-FXT6D4 | Cat. No. 800H-FRXT6D4 |

Figure 2-6 *2-position red push-pull and push-pull/twist release devices, nonilluminated. Courtesy of Rockwell Automation, Inc. For detailed specifications, see Allen Bradley at www.ab.com.*

The buttons in Fig. 2-6 are mechanically maintained. Typical use is for emergency stop or start, as in turning off all power in an IT Equipment Room (EPO) or starting the halon system in the event of a fire. For detailed specifications including a complete list of contacts, see Allen Bradley at www.ab.com.

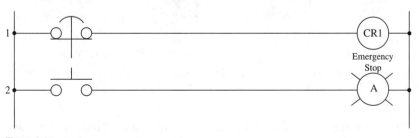

Figure 2-6a

When the push button in Fig. 2-6a is pressed, the relay CR1 loses power and the pilot light is energized.

36   Chapter Two

Note: A jumbo or large legend plate is recommended, if space allows.

Illuminated 2-Position Push-Pull   *2-Position Push-Pull*   *Illuminated 2-Position Push-Pull/Twist*   *Illuminated 2-Position Push-Pull/Twist*
Cat. No. 800T-FXP16RA1   Cat. No. 800T-FXJEP16RA1   Cat. No. 800T-FXTP16RA1   Cat. No. 800H-FRXTP16RA1

Figure 2-7 2-position red push-pull and push-pull/twist release devices, illuminated. *Courtesy of Rockwell Automation, Inc.*

The buttons in Fig. 2-7 are mechanically maintained and illuminated. Typical use is for emergency stop or start, as in turning off all power in an IT Equipment Room (EPO) or starting the halon system in the event of a fire. For detailed specifications, see Allen Bradley at www.ab.com.

Figure 2-7a

When the push button in Fig. 2-7a is pressed, the relay CR1 loses power and the pilot light is energized.

Flush Head Unit   **Extended Head Unit**   **Booted Unit**   **Bootless Flush Head Unit**
Cat. No. 800T-A1A   Cat. No. 800T-86A   Cat. No. 800H-R2A   Cat. No. 800H-AR1A

Figure 2-8 Momentary contact push-button devices, nonilluminated. *Courtesy of Rockwell Automation, Inc.*

The push buttons in Fig. 2-8 can have multiple contacts. It is recommended that this push button have a maximum of four 2 × 2 blocks. This can give it a total of 8 contacts. If more than this is needed, then you should consider using a relay. These operators can use the contacts in Fig. 2-3. For detailed specifications, see Allen Bradley at www.ab.com.

Input Devices   37

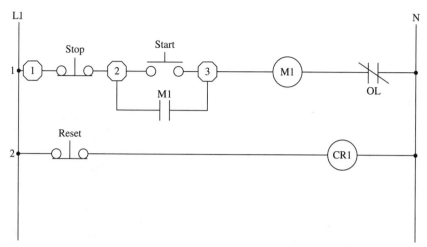

Figure 2-8a

Typically the Stop button will be an extended head unit red. The Start button will be a flush head unit green. The Reset button will be an extended head unit black.

Shown in Fig. 2-8a:

Line 1 The Stop button is closed. When the Start button is pressed, the M1 motor starter is energized and the M1 contact closes, and the motor starts and runs until the Stop button is pressed.

Line 2 When the Reset button is pressed, the relay CR1 loses power and resets something not shown.

Extended Head without Guard
Cat. No. 800T-PB16R

Extended Head without Guard
Cat. No. 800H-PRB16R

Figure 2-9 Momentary contact push-button devices, illuminated. *Courtesy of Rockwell Automation, Inc.*

The illuminated push buttons in Fig. 2-9 can have two contact blocks. This can give it a total of 4 contacts. If more than this is needed, then you should consider using a relay. These operators can use the contacts in Fig. 2-3. For detailed specifications, see Allen Bradley at www.ab.com.

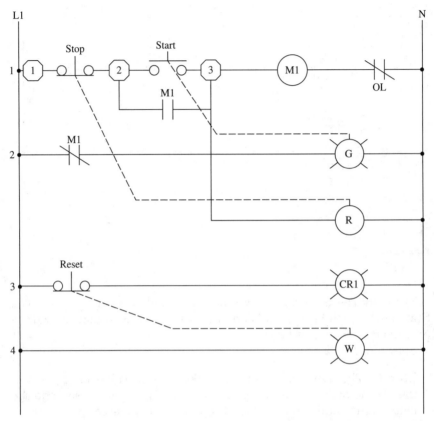

Figure 2-9a The red Stop pilot light will be off, and the green Start pilot light will be on.

Shown in Fig. 2-9a:

Line 1 Stop button is closed. When the Start button is pressed, the M1 motor starter is energized, and the M1 contact closes, the red pilot light will be on, and the motor starts and runs until the Stop button is pressed.

Line 2 M1 contact will open and the green pilot light will be off.

Line 3 When the Reset button is pressed, the relay CR1 loses power and resets something not shown.

Line 4 White pilot light is on.

The push buttons in Fig. 2-10 have molded legends and can have multiple contacts. It is recommended that this push button have a maximum of four 2 × 2 blocks. This can give a total of 8 contacts. If more than this is needed, then you should consider using a relay. These operators can use the contacts in Fig. 2-3. For detailed specifications, see Allen Bradley at www.ab.com.

Cat. No. 800T-A00 with Cat. No. 800T-LC103W installed   Cat. No. 800H-BR00 with Cat. No. 800T-LC604 installed

Figure 2-10 Momentary contact push-button devices, nonilluminated—with two-color molded legend caps. *Courtesy of Rockwell Automation, Inc.*

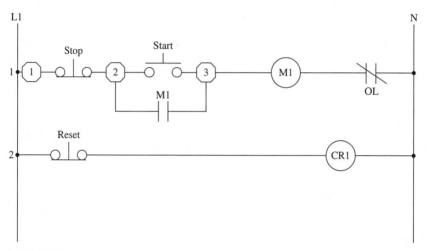

Figure 2-10a

Line 1 The Stop button is closed. When the Start button is pressed, the M1 motor starter is energized and the M1 contact closes, and the motor starts and runs until the Stop button is pressed.

Line 2 When the Reset button is pressed, the relay CR1 loses power and resets something not shown.

Standard Knob Operator   *Knob Lever Operator*   *Standard Knob Operator*
Cat. No. 800T-H2A   Cat. No. 800T-H17A   Cat. No. 800H-HR2A

Figure 2-11 2-position Selector switch devices, nonilluminated. *Courtesy of Rockwell Automation, Inc.*

The 2-position Selector switch operators in Fig. 2-11 can have multiple contacts. It is recommended that this operator have a maximum of four 2 × 2 blocks. This can give a total of 8 contacts, which can be Normally Open or Normally Closed. If more than this is needed, then you should consider using a relay. Typical use is for On/Off operators. These operators can use the contacts in Fig. 2-3. For detailed specifications, see Allen Bradley at www.ab.com.

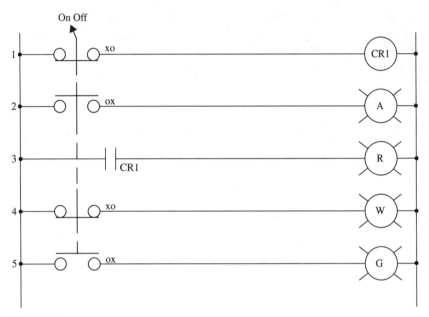

Figure 2-11a

Shown in Fig. 2-11a:

Line 1 Selector switch is in the On position as indicated by the XO relay CR1 being energized.

Line 2 Selector switch is in the Off position as indicated by the OX relay; amber pilot light is energized.

Line 3 If relay CR1 is energized, then red pilot light is energized.

Line 4 Selector switch is in the On position as indicated by the XO relay; white pilot is energized.

Line 5 Selector switch is in the Off position as indicated by OX relay; green pilot light is energized.

The 2-position keyed Selector switch operator in Fig. 2-12 can have multiple contacts. It is recommended that it have a maximum of four

2-Position Cylinder Lock Operator
Cat. No. 800T-H33A

Figure 2-12 2-position Selector switch device, nonilluminated (800T only). *Courtesy of Rockwell Automation, Inc.*

2 × 2 blocks. This can give a total of 8 contacts, which can be Normally Open or Normally Closed. If more than this is needed, then you should consider using a relay. Typical use is for On/Off operators. These operators can use the contacts in Fig. 2-3. For detailed specifications, see Allen Bradley at www.ab.com.

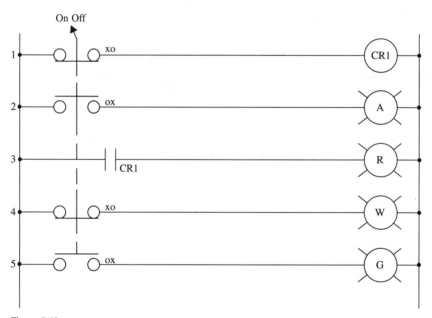

Figure 2-12a

Shown in Fig. 2-12a:

Line 1 Keyed Selector switch is in the On position as indicated by XO relay; CR1 is energized.

Line 2 Keyed Selector switch is in the Off position as indicated by OX relay; amber pilot light is energized.

Line 3 If relay CR1 is energized, then the red pilot light is energized.

Line 4 Keyed Selector switch is in the On position as indicated by XO relay; white pilot light is energized.

Line 5 Keyed Selector switch is in the Off position as indicated by OX relay; green pilot light is energized.

Standard Knob Operator
Cat. No. 800T-J2A

Knob Lever Operator
Cat. No. 800T-J17A

Standard Knob Operator
Cat. No. 800H-JR2A

Figure 2-13 3-position Selector switch devices, nonilluminated. *Courtesy of Rockwell Automation, Inc.*

The 3-position Selector switch operators such as these can have multiple contacts (see Fig. 2-13). It is recommended that this operator have a maximum of four 2 × 2 blocks. This can give it a total of 8 contacts, which can be Normally Open or Normally Closed. If more than this is needed, then you should consider using a relay. Typical use is for Hand/Off/Auto operators. These operators can use the contacts in Fig. 2-3. For detailed specifications, see Allen Bradley at www.ab.com.

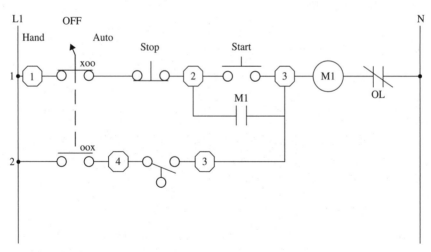

Figure 2-13a

Line 1 The 3-position Selector switch is in the Hand position as indicated by the XOO, and the Stop button is closed. When the Start button is pressed, the motor starter will be energized and the

M1 contact will close. The motor will run until the Stop button is depressed or the select switch position is changed.

Line 2 3-position Selector switch is in the Auto position as indicated by the OOX, and the float switch is closed. The motor will run until the liquid level drops and opens the float switch.

*3-Position Cylinder Lock Operator
Cat. No. 800T-J41A*

Figure 2-13b 3-position Selector switch device, nonilluminated (800T only). *Courtesy of Rockwell Automation, Inc.*

The 3-position keyed Selector switch operator in Fig. 2-13b can have multiple contacts. It is recommended that it have a maximum of four 2 × 2 blocks. This can give it a total of 8 contacts, which can be Normally Open or Normally Closed. If more than this is needed, then you should consider using a relay. Typical use is for Hand/Off/Auto operator. These operators can use the contacts in Fig. 2-3. For detailed specifications, see Allen Bradley at www.ab.com.

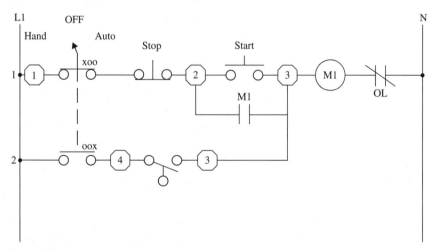

Figure 2-13c

Shown in Fig. 2-13c:

Line 1 3-position keyed Selector is in the Hand position as indicated by the XOO, and the Stop button is closed. When the Start button is pressed, the motor starter will be energized and the M1 contact will close. The motor will run until the Stop button is depressed or the select switch position is changed.

Line 2   3-position keyed Selector is in the Auto position as indicated by the OOX, and the float switch is closed. The motor will run until the liquid level drops and opens the float switch.

Standard Knob Operator   *Knob Lever Operator*   *Standard Knob Operator*
Cat. No. 800T-N2KN4B   *Cat. No. 800T-N17KN4B*   *Cat. No. 800H-NR2KF4AAXX*

Figure 2-14   4-position Selector switch devices, nonilluminated. *Courtesy of Rockwell Automation, Inc.*

The 4-position Selector switch operators in Fig. 2-14 can have multiple contacts. It is recommended that this operator have a maximum of four 2 × 2 blocks. This can give it a total of 8 contacts, which can be Normally Open or Normally Closed. If more than this is needed, then consider using a relay. Typical use is for Hand operation 1 operation 2 Auto operator. These operators can use the contacts in Fig. 2-3. For detailed specifications, see Allen Bradley at www.ab.com.

Shown in Fig. 2-14a:

Line 1   4-position Selector is in the Hand position as indicated by the XOOO, and the Stop button is closed. When the Start button is pressed, the motor starter will be energized, the M1 contact will close, and pump 1 will run until the Stop button is depressed or the select switch position is changed.

Line 2   4-position Selector is in the pump 1 or Auto position as indicated by the OXOO and the OOOX, and the float switch is closed. Pump 1 will run until the liquid level drops and opens the float switch.

Line 3   4-position Selector is in the Hand position as indicated by the XOOO, and the Stop button is closed. When the Start button is pressed, the motor starter will be energized, the M2 contact will close, and pump 2 will run until the Stop button is depressed or the select switch position is changed.

Line 4   4-position Selector is in the pump 2 or Auto position as indicated by the OOXO and the OOOX, and the float switch is closed. Pump 2 will run until the liquid level drops and opens the float switch.

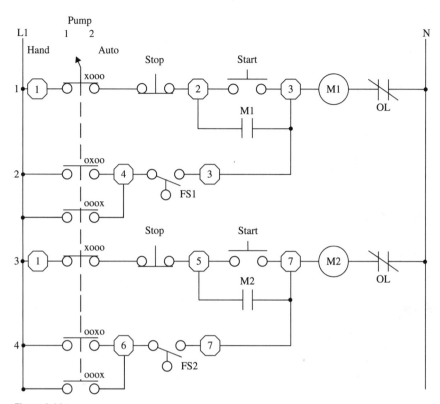

Figure 2-14a

Cylinder Lock Operator
Cat. No. 800T-N32KF48

Figure 2-15 4-position Selector switch device, nonilluminated (800T only). *Courtesy of Rockwell Automation, Inc.*

The 4-position keyed Selector switch operator in Fig. 2-15 can have multiple contacts. It is recommended that it have a maximum of four 2 × 2 blocks. This can give it a total of 8 contacts, which can be Normally Open or Normally Closed. If more than this is needed, then you should consider using a relay. Typical use is for Hand operation 1 operation 2 Auto operator. These operators can use the contacts in Fig. 2-3. For detailed specifications, see Allen Bradley at www.ab.com.

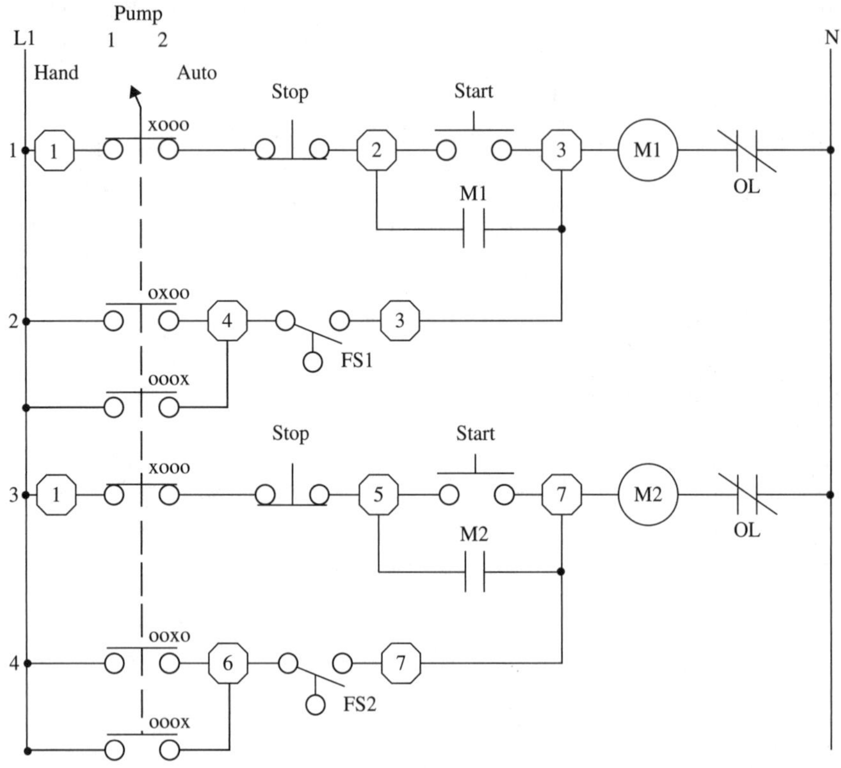

Figure 2-15a

Shown in Fig. 2-15a:

Line 1 4-position keyed Selector is in the Hand position as indicated by the XOOO, and the Stop button is closed. When the Start button is pressed, the motor starter will be energized, the M1 contact will close, and pump 1 will run until the Stop button is depressed or the select switch position is changed.

Line 2 4-position keyed Selector is in pump 1 or Auto position as indicated by the OXOO and the OOOX, and the float switch is closed. Pump 1 will run until the liquid level drops and opens the float switch.

Line 3 4-position keyed Selector is in the Hand position as indicated by the XOOO, and the Stop button is closed. When the Start button is pressed, the motor starter will be energized, the M2 contact will close, and pump 2 will run until the Stop button is depressed or the select switch position is changed.

Line 4 4-position keyed Selector is in pump 2 or Auto position as indicated by the OOXO and the OOOX, and the float switch is closed. Pump 2 will run until the liquid level drops and opens the float switch.

Standard Knob Operator   *Knob Lever Operator*
Cat. No. 800T-16HR2KB6AX   Cat. No. 800H-16HRR17KB6AX

Figure 2-16 2-position knob/lever type Selector switch devices, illuminated. *Courtesy of Rockwell Automation, Inc.*

The 2-position illuminated Selector switch operators in Fig. 2-16 can be spring-return or maintained-position. This operator can have multiple contacts; it is recommended that it have a maximum of four 2 × 2 blocks. This can give it a total of 8 contacts, which can be Normally Open or Normally Closed. If more than this is needed, then you should consider using a relay. Typical use is for On/Off operators. These operators can use the contacts in Fig. 2-3. For detailed specifications, see Allen Bradley at www.ab.com.

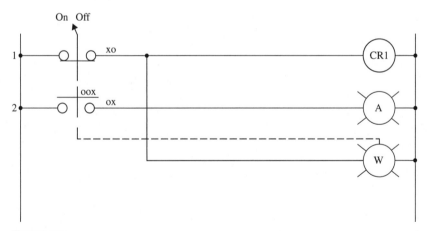

Figure 2-16a

Shown in Fig. 2-16a:

Line 1 2-position Selector switch is in the On position as indicated by the XO; CR1 will be energized and the white pilot light will be on.

Line 2 2-position Selector switch is in the Off position as indicated by the OX; the amber pilot light will be on.

Standard Knob Operator
Cat. No. 800T-16JR2KB7AX

Standard Knob Operator
Cat. No. 800H-16JRR2KB7AX

Figure 2-17 3-position knob/lever type Selector switch devices, illuminated. *Courtesy of Rockwell Automation, Inc.*

The 3-position lever illuminated Selector switch operators in Fig. 2-17 can be spring-return or maintain-position. This operator can have multiple contacts. It is recommended that it have two blocks. This can give it a total of 4 contacts, which can be Normally Open or Normally Closed. If more than this is needed, then you should consider using a relay. Typical use is for Hand/Off/Auto operator. These operators can use the contacts in Fig. 2-3. For detailed specifications, see Allen Bradley at www.ab.com.

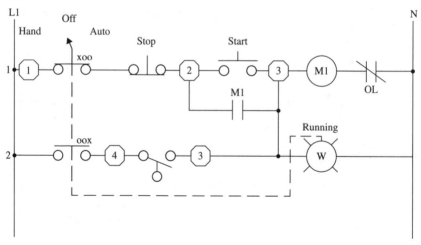

Figure 2-17a

Shown in Fig. 2-17a:

Line 1 3-position keyed Selector is in the Hand position as indicated by the XOO, and the Stop button is closed. When the Start button is pressed, the motor starter will be energized, the M1 contact will close, and the motor will run until the Stop button is depressed or the select switch position is changed.

Line 2 3-position keyed Selector is in the Auto position as indicated by the OOX, and the float switch is closed. The motor will run and the white pilot light will be on until the liquid level drops and opens the float switch.

Note: A jumbo or large legend plate is recommended, if space allows.

3-Position Push-Pull
Cat. No. 800T-FXM6A7

3-Position Push-Pull
Cat. No. 800H-FRXM6A7

Figure 2-18 3-position push-pull devices, nonilluminated. *Courtesy of Rockwell Automation, Inc.*

In the devices in Fig. 2-18, the center position is maintained. The top and bottom positions can be momentary or maintained. The button has one Normally Closed contact and one Normally Closed late break contact. For detailed specifications, see Allen Bradley at www.ab.com.

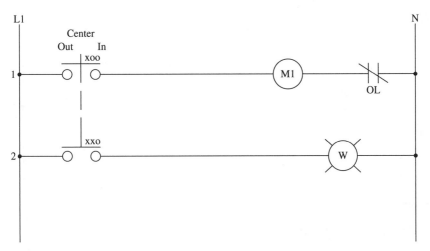

Figure 2-18*a*

Shown in Fig. 2-18*a*:

Line 1 3-position push-pull device is in the Out position as indicated by the XOO; the motor starter will be energized, and the motor will run.

Line 2   3-position push-pull device is in the Out or center position as indicated by the XXO. The white pilot light will be on.

Illuminated 3-Position Push-Pull
Cat. No. 800T-FXMP16RA7

Illuminated 3-Position Push-Pull
Cat. No. 800H-FRXMP16A7

Figure 2-19   3-position push-pull devices, illuminated. *Courtesy of Rockwell Automation, Inc.*

In the devices in Fig. 2-19, the center position is maintained. The top and bottom positions can be momentary or maintained. The button has one Normally Closed contact and one Normally Closed late break contact and has an internal pilot light. For detailed specifications, see Allen Bradley at www.ab.com.

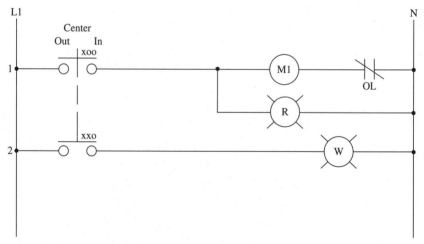

Figure 2-19a

Shown in Fig. 2-19a:

Line 1   3-position push-pull device is in the Out position as indicated by the XOO. The motor starter will be energized, the motor will run, and the red pilot light will be on.

Line 2   3-position push-pull device is in the Out or center position as indicated by the XXO. The white pilot light will be on.

Input Devices   51

Cat. No. 800T-U24   Cat. No. 800H-UR4

Figure 2-20 Potentiometer devices. *Courtesy of Rockwell Automation, Inc.*

Potentiometers are used any time a variable resistance is needed (see Fig. 2-20). Caution should be taken to ensure that the potentiometer is not overloaded as this will cause damage. Typical use is to set the time delay of a relay. For detailed specifications, see Allen Bradley at www.ab.com.

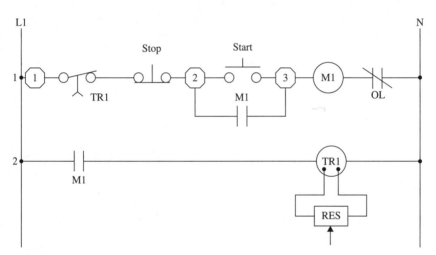

Figure 2-20a

Shown in Fig. 2-20a:

Line 1 Terminal 1, then Normally Closed time contact from time delay relay TR1. The contact will open after a preset time when the TR1 is energized. Normally Closed Stop button, terminal 2 Normally Open Start button. Down one line, Normally Open M1 contact. Motor starter M1 with Normally Closed overload contact. When the Start button is pressed, then the M1 motor will be energized and the motor will run until the Stop button or the relay TR1 times out.

Line 2 Normally Open contact of motor starter M1. Time delay relay TR1 with adjustable time delay by use of the potentiometer. When the motor starter is energized, the contact M1 will close and start the time cycle set by the potentiometer.

Cat. No. 800T-FA22A   Cat. No. 800T-FB16A   Cat. No. 800T-FC16F   Cat. No. 800H-CRA22A

Figure 2-21 Mechanically interlocked maintained push-button devices. *Courtesy of Rockwell Automation, Inc.*

In Fig. 2-21, the push buttons are maintained; one contact is open and one contact is closed. This operator can have a maximum of two Normally Open and two Normally Closed. Typical use is Start-Stop station. For detailed specifications, see Allen Bradley at www.ab.com.

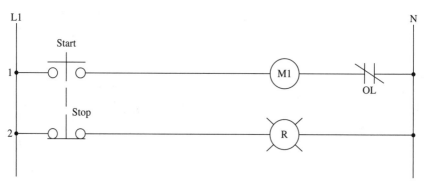

Figure 2-21a

Shown in Fig. 2-21a:

Line 1 Normally Open Start button motor start M1 with Normally Closed overload relay. When the Start button is pressed, the motor will run and the red pilot light will be off.

Line 2 Normally Closed Stop button. When it is pressed, the motor will stop and the red pilot light will be on.

Input Devices   53

Color Cap Only With Lamps   *Complete Unit*
Cat. No. 800T-N249   *Cat. No. 800T-PC416*

Figure 2-22 Cluster pilot light devices (800T only).
Courtesy of Rockwell Automation, Inc.

The cluster pilot lights in Fig. 2-22 can have 2, 3, or 4 lights with different-color lenses. For detailed specifications, see Allen Bradley at www.ab.com.

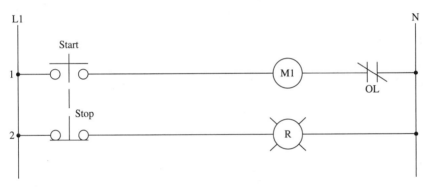

Figure 2-22a

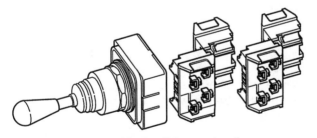

Figure 2-23 *Courtesy of Rockwell Automation, Inc.*

In Fig. 2-23 there is a 4-position toggle switch. This operator can have 1, 2, 3, or 4 positions. The operator can be momentary or maintained-position; in Fig. 2-23 it is a maintained operator. This operator can have multiple contacts. It is recommended that it have a maximum of four 2 × 2 blocks. This can give it a total of 8 contacts, which can be Normally Open or Normally Closed. If more than this is needed, then consider using a relay. Typical use is for Hand operation 1 operation 2 Auto operator. These operators can use the contacts in Fig. 2-3. For detailed specifications, see Allen Bradley at www.ab.com.

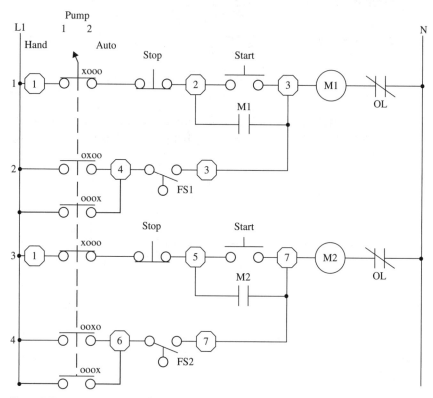

Figure 2-23a

Line 1 4-position toggle Selector is in the Hand position as indicated by the XOOO, and the Stop button is closed. When the Start button is pressed, the motor starter will be energized and the M1 contact will close. Pump 1 will run until the Stop button is depressed or the select switch position is changed.

Line 2 4-position toggle selector is in the pump 1 or Auto position as indicated by the OXOO and the OOOX, and the float switch is closed. Pump 1 will run until the liquid level drops and opens the float switch.

Line 3 4-position toggle selector is in the Hand position as indicated by the XOOO, and the Stop button is closed. When the Start button is pressed, the motor starter will be energized and the M2 contact will close. Pump 2 will run until the Stop button is depressed or the select switch position is changed.

Line 4 4-position toggle selector is in the pump 2 or Auto position as indicated by the OOXO and the OOOX, and the float switch is closed. Pump 2 will run until the liquid level drops and opens the float switch.

Figure 2-24 Selector push-button device (800T only). *Courtesy of Rockwell Automation, Inc.*

Selector Push Button
Cat. No. 800T-K2AAXX

The 2-position Selector switch with push-button operator in Fig. 2-24 can have multiple contacts. It is recommended that it have a maximum of two 1 × 1 blocks. This can give it a total of 4 contacts, which can be two Normally Open and two Normally Closed. The sleeve is the Selector switch; it can be moved left and right. Typical use is for the run jog operator. These operators can use the contacts in Fig. 2-3. For detailed specifications, see Allen Bradley at www.ab.com.

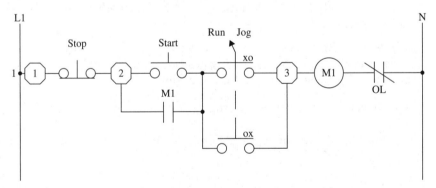

Figure 2-24a

Shown in Fig. 2-24a:

Line 1 Terminal 1, Normally Closed push button, terminal 2. Down one line, Normally Open M1 contact. Down one line, contact of the 2-position Selector switch closed in the jog position indicated by the OX. Next on line 1, contact of the 2-position Selector switch closed in the run position indicated by the XO. Terminal 3, M1 motor starter with Normally Closed overload contact. With the run jog in the run position, pressing the Start button will energize the M1 motor starter and close the M1 contact. The Motor will run until the Stop button is pressed or the Selector switch is changed. With the run jog Selector switch in the jog position, pressing the Start button will energize motor starter M1 as long as the button is pressed.

Cylinder Lock Push Button   *Mushroom Style Cylinder Lock*
Cat. No. 800T-E15A   Cat. No. 800T-E15M6A

Figure 2-25 Cylinder lock push-button devices (800T only). *Courtesy of Rockwell Automation, Inc.*

The operators in Fig. 2-25 can be locked in the depressed position. The button is available in three modes:

Spring bolt Lock can be set with a key when button is in the Out position. Button will lock when depressed. Key is removable in any position. One Normally Open and one Normally Closed lock position in.

Dead bolt A Button can only be operated with the key in the lock. Key is removable in locked position only. One Normally Open and one Normally Closed lock position in or out.

Dead bolt B Button can be operated with or without the key inserted in the lock. Key is removable in any position. One Normally Open and one Normally Closed lock position in or out.

Typical use is for emergency stop on motor circuit or on an EPO circuit.

For detailed specifications, see Allen Bradley at www.ab.com.

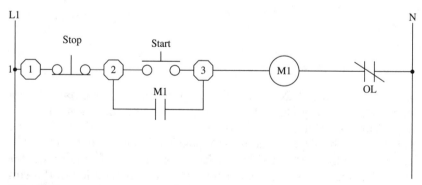

Figure 2-25a

Shown in Fig. 2-25a:

Line 1 Terminal 1, then Normally Closed Stop button, terminal 2 Normally Open Start button. Down one line, Normally Open M1 contact. Motor starter M1 with Normally Closed overload contact. When the Start button is pressed, then the M1 motor will be energized and the motor will run until the Stop button is pressed.

Padlocking Mushroom Button *Padlocking Jumbo Mushroom Button*
Cat. No. 800T-D6QA *Cat. No. 800T-D6LQA*

Figure 2-26 Momentary padlocking mushroom head devices (800T only). *Courtesy of Rockwell Automation, Inc.*

The mushroom push buttons in Fig. 2-26 can have one Normally Open or one Normally Closed or one Normally Open and one Normally Closed contact. This allows locking in the depressed position. A mushroom push button will hold Normally Closed contacts open, but might not hold Normally Open contacts closed. For detailed specifications, see Allen Bradley at www.ab.com.

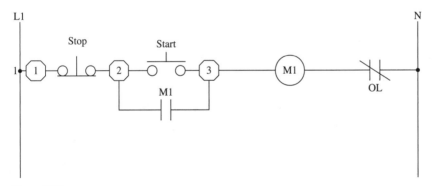

Figure 2-26a

Shown in Fig. 2-26a:

Line 1 Terminal 1, then Normally Closed Stop button, terminal 2 Normally Open Start button. Down one line, Normally Open M1 contact. Motor starter M1 with Normally Closed overload contact. When the Start button is pressed, then the M1 motor will be energized and the motor will run until the Stop button is pressed.

Flip Lever Operator *Flip Lever Operator*
Cat. No. 800H-WK4B   Cat. No. 800H-WK61B

Figure 2-27 Momentary contact flip lever devices—800H type 414X only. *Courtesy of Rockwell Automation, Inc.*

The push buttons in Fig. 2-27 can have one Normally Open or one Normally Closed or two Normally Open or two Normally Closed contacts. This prevents accidental use of the push button. Typical use is for Stop and Start buttons. For detailed specifications, see Allen Bradley at www.ab.com.

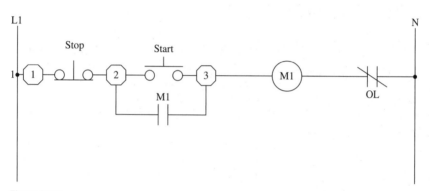

Figure 2-27a

Shown in Fig. 2-27a:

Line 1 Terminal 1, then Normally Closed stop button, terminal 2 Normally Open Start button. Down one line, Normally Open M1 contact. Motor starter M1 with Normally Closed overload contact. When the Start button is pressed, then the M1 motor will be energized and the motor will run until the Stop button is pressed.

Figure 2-28 Momentary wobble stick push-button device—800T type 13 only. *Courtesy of Rockwell Automation, Inc.*

Wobble Stick Unit
Cat. No. 800T-M1B

The momentary wobble device in Fig. 2-28 will open a Normally Closed contact and close a Normally Open contact. This operator is available with one Normally Open contact or with one Normally Open and one Normally Closed contact or with two Normally Open and two Normally Closed contacts. For detailed specifications, see Allen Bradley at www.ab.com.

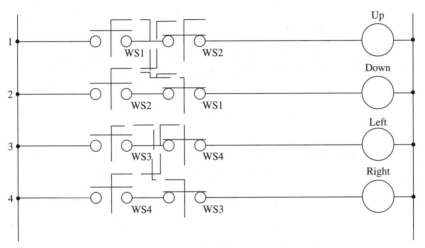

Figure 2-28a

Shown in Fig. 2-28a:

Line 1 Normally Open contact of wobble stick 1, Normally Closed contact of wobble stick 2, up relay.

Line 2 Normally Open contact of wobble stick 2, Normally Closed contact of wobble stick 1, down relay.

Line 3 Normally Open contact of wobble stick 3, Normally Closed contact of wobble stick 4, left relay.

Line 4 Normally Open contact of wobble stick 4, Normally Closed contact of wobble stick 3, right relay

When the up wobble stick 1 is moved, the contact for the up relay will close and the contact for the down relay will open, preventing the up and down relays from being used at the same time.

When the down wobble stick 2 is moved, the contact for the down relay will close and the contact for the up relay will open, preventing the up and down relays from being used at the same time.

When the left wobble stick 3 is moved, the contact for the left relay will close and the contact for the right relay will open, preventing the left and right relays from being used at the same time.

Chapter Two

When the right wobble stick 4 is moved, the contact for the right relay will close and the contact for the left relay will open, preventing the left and right relays from being used at the same time.

For detailed specifications on the above items, see the Allen Bradley Catalog at www.ab.com/en/epub/catalogs/12768/229240/229244/2531083/tab4.html.

Automatic Float Switch

Float switches provide automatic control for motors that pump liquids from a sump or into a tank (Fig. 2-29a, b, and c). The switch must be

- Liquid Level Sensitivity From 2 to 5 Inches
- 0.6 to 3.8 lbs Switch Operating Force
- 2 or 3 Pole Contact Configurations
- Tank to Sump Convertibility
- NEMA A600 and NEMA N300 Contact Ratings
- Type 1, Type 4, and Type 7 & 9 Enclosures

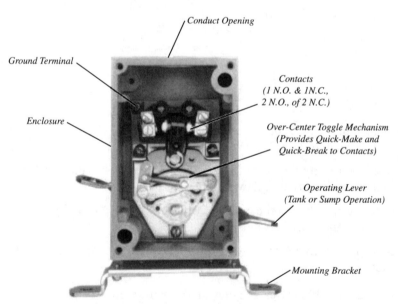

Figure 2-29a Bulletin 840 Style A shown with cover removed. *Courtesy of Rockwell Automation, Inc.*

Figure 2-29b Bulletin 840 Type 7 & 9. *Courtesy of Rockwell Automation, Inc.*

Input Devices   61

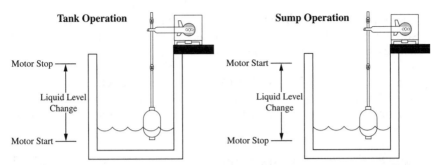

Figure 2-29c Tank and sump operation. *Courtesy of Rockwell Automation, Inc.*

installed above the tank or sump, and the float must be in the liquid for the float switch to operate. For detailed specifications, see Allen Bradley at www.ab.com.

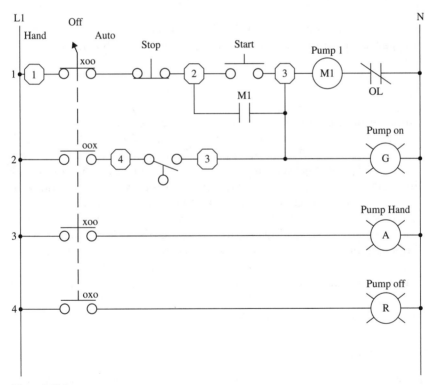

Figure 2-29d

Shown in Fig. 2-29d:

Line 1   Terminal 1, contact of Hand/Off/Auto Selector switch closed in the Hand position as indicated by the XOO, Normally Closed Stop button, terminal 2. Down one line, Normally Open contact of M1 starter. Normally Open Start push button, terminal 3, motor starter with Normally Closed overload contact.

Line 2 Contact of Hand/Off/Auto Selector switch closed in the Auto position as indicated by the OOX, terminal 4, Normally Open contact of the float switch, connection to starter circuit, green pilot light.

Line 3 Contact of Hand/Off/Auto Selector switch closed in the Hand position as indicated by the XOO, amber pilot light.

Line 4 Contact of Hand/Off/Auto Selector switch closed in the Off position as indicated by the OXO.

Figure 2-29d is an example of a sump operation:

Hand In the Hand position the Stop and Start buttons can be used to operate the pump, and the amber pilot light will be on.

Off In the Off position the circuit is inactive and the red pilot light will be on.

Auto A float operator assembly is attached to the float switch by a rod, chain, or cable. The float switch is actuated based on the location of the float in the liquid. The float switch contacts are closed when the float forces the operating lever to the up position. As the liquid level rises, the float and operating lever move upward. When the float reaches a preset high level, the float switch contacts close when the select switch is in the Auto position, activating the circuit and starting the motor and turning on the green pilot light. As the liquid level falls, the float and operating lever move downward. When the float reaches a preset low level, the float switch contacts open, deactivating the circuit and stopping the motor.

Tank operation is the opposite of sump operation.

Pressure Switch

Pressure switches provide automatic control for motors that pump pressure into a line (Fig. 2-30a, b, and c). The switch must be installed in the line to be monitored. There are two circuits, with one Normally Open and one Normally Closed contact per circuit.

For detailed specifications, see Allen Bradley at www.ab.com.

Input Devices

Bulletin 836T Pressure Controls

- Operating ranges from 30 in. Hg vacuum...5000 psi
- Independently adjustable range and differential
- Copper alloy and stainless steel bellows
- 2- and 4-Circuit contact block
- Pressure difference controls available
- 1/4 in. and 3/8 in. N.P.T. and O-ring straight thread connections
- Type 4 & 13 and Type 7 & 9 and 4 & 13 combination enclosures

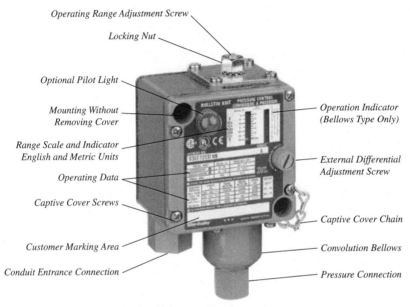

Figure 2-30a *Courtesy of Rockwell Automation, Inc.*

Figure 2-30b *Courtesy of Rockwell Automation, Inc.*

64 Chapter Two

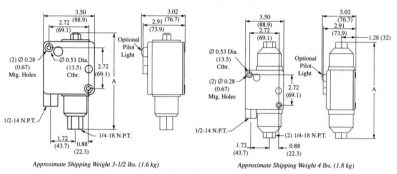

Figure 2-30c Type 4 & 13 (bellows). *Courtesy of Rockwell Automation, Inc.*

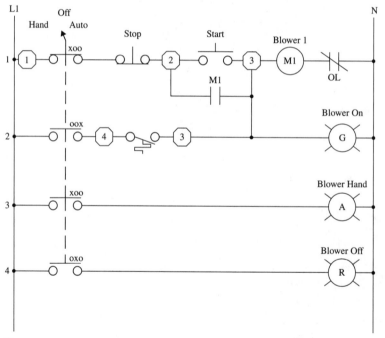

Figure 2-30d

Shown in Fig. 2-30d:

Line 1 Terminal 1, contact of Hand/Off/Auto Selector switch closed in the Hand position as indicated by the XOO, Normally Closed Stop button, terminal 2. Down one line, Normally Open contact of M1 starter. Normally Open Start push button, terminal 3, motor starter with Normally Closed overload contact.

Line 2 Contact of Hand/Off/Auto Selector switch closed in the Auto position as indicated by the OOX, terminal 4, Normally Closed contact of the pressure switch, connection to starter circuit, green pilot light.

Line 3 Contact of Hand/Off/Auto Selector switch closed in the Hand position as indicated by the XOO, amber pilot light.

Line 4 Contact of Hand/Off/Auto Selector switch closed in the Off position as indicated by the OXO.

Figure 2-30d is an air compressor:

Hand In the Hand position, the Stop and Start buttons can be used to operate the compressor, and the amber pilot light will be on.

Off In the Off position the circuit is inactive and the red pilot light will be on.

Auto The pressure switch is actuated based on the air pressure in the line being monitored. The pressure switch contacts are closed; when the air pressure level decreases, the pressure in bellows decreases. When the pressure reaches a preset low level, the pressure switch contacts close with the select switch in the Auto position activating the circuit and starting the motor and turning on the green pilot light. As the pressure level increases, the pressure in the bellows increases. When the air pressure reaches a preset high level, the pressure switch contacts open, deactivating the circuit and stopping the motor.

Temperature Switch

Bulletin 837 temperature controls are heavy-duty control circuit devices used in industrial applications where the temperature must be maintained within preset limits (Fig. 2-31a, b, and c). These devices use a vapor pressure technology to sense changes in temperature. The pressure change is transmitted to the bellows through a bulb and capillary tube. Pressure in the system changes in proportion to the temperature of the bulb. Most styles make use of bulbs and capillaries filled with a temperature-responsive liquid for detecting temperature changes.

For detailed specifications, see Allen Bradley at www.ab.com.

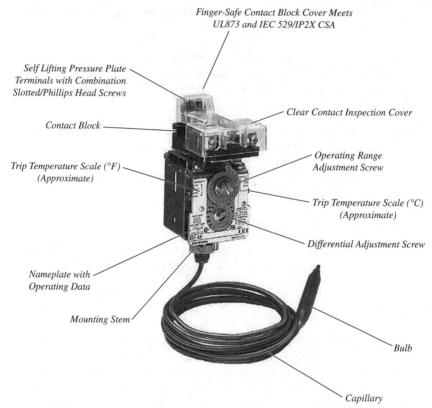

Figure 2-31a Bulletin 837 Bulb and Capillary Type without enclosure. *Courtesy of Rockwell Automation, Inc.*

Figure 2-31b *Courtesy of Rockwell Automation, Inc.*

Input Devices 67

Dimensions in inches (millimeters). Dimensions are not intended to be used for manufacturing purposes.

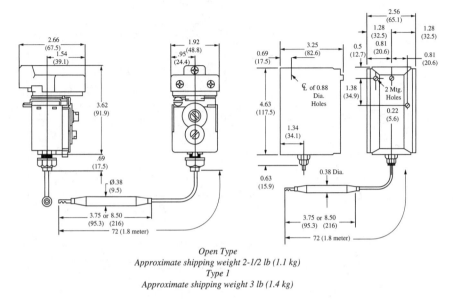

Open Type
Approximate shipping weight 2-1/2 lb (1.1 kg)
Type 1
Approximate shipping weight 3 lb (1.4 kg)

Figure 2-31c Remote bulb and capillary type. *Courtesy of Rockwell Automation, Inc.*

Shown in Fig. 2-31*d*:

Line 1 Terminal 1, contact of Hand/Off/Auto Selector switch closed in the Hand position as indicated by the XOO, Normally Closed Stop button, terminal 2. Down one line, Normally Open contact of M1 starter. Normally Open Start push button, terminal 3, motor starter with Normally Closed overload contact.

Line 2 Contact of Hand/Off/Auto Selector switch closed in the Auto position as indicated by the OOX, terminal 4, Normally Closed contact of the temperature switch, connection to starter circuit, green pilot light.

Line 3 Contact of Hand/Off/Auto Selector switch closed in the Hand position as indicated by the XOO, amber pilot light.

Line 4 Contact of Hand/Off/Auto Selector switch closed in the Off position as indicated by the OXO.

Figure 2-31*d* is a blower circuit:

Hand In the Hand position the Stop and Start buttons can be used to operate the blower, and the amber pilot light will be on.

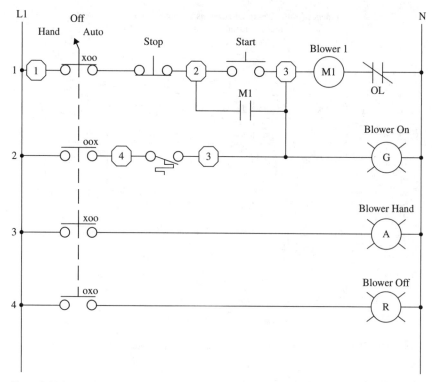

Figure 2-31d

Off In the Off position the circuit is inactive and the red pilot light will be on.

Auto The temperature switch is actuated based on the temperature in the duct being monitored. The temperature switch contacts are closed; when the temperature level raises, the pressure in bellows increases. When the temperature reaches a preset level, the temperature switch contacts close with the select switch in the Auto position, activating the circuit, starting the motor, and turning on the green pilot light. As the temperature level decreases, the pressure in the bellows decreases. When the temperature reaches a preset low level, the pressure switch contacts open, deactivating the circuit and stopping the motor.

Photoelectric Sensor

There are four basic components to any photosensor (Fig. 2-32a, b, c): light source, light detector, lenses, and output switching device. There are basically two types of photosensors: LED and photoelectric.

Input Devices   69

Figure 2-32a RightSight DC model with short 18-mm base. *Courtesy of Rockwell Automation, Inc.*

Figure 2-32b *Courtesy of Rockwell Automation, Inc.*

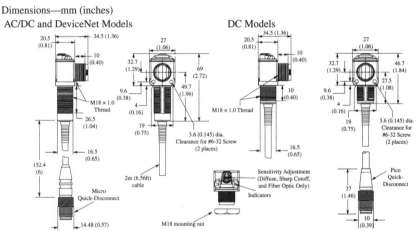

Figure 2-32c *Courtesy of Rockwell Automation, Inc. For detailed specifications, see Allen Bradley at www.ab.com.*

Shown in Fig. 2-32d:

Line 1 Terminal 1, contact of Hand/Off/Auto Selector switch closed in the hand position as indicated by the XOO, Normally Closed Stop button, terminal 2. Down one line, Normally Open Contact of M1 starter. Normally Open Start push button, terminal 3, motor starter with Normally Closed overload contact.

Line 2 Contact of control relay CR1.

Line 3 Output of photosensor PS1, solid-state relay SSR1.

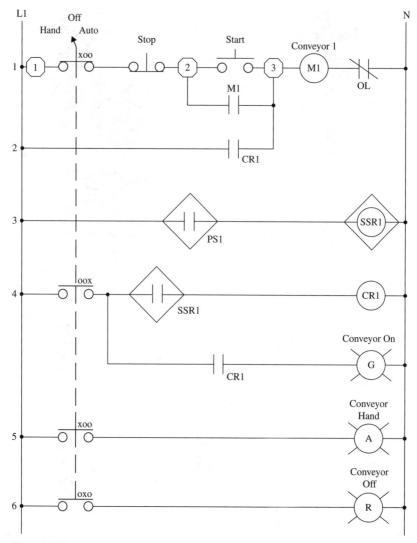

Figure 2-32d

Line 4 Contact of Hand/Off/Auto Selector switch closed in the Auto position as indicated by the OOX, Normally Closed contact of the solid-state relay SSR1. Down one line, green pilot light.

Line 5 Contact of Hand/Off/Auto Selector switch closed in the Hand position as indicated by the XOO, amber pilot light.

Line 6 Contact of Hand/Off/Auto Selector switch closed in the Off position as indicated by the OXO.

Figure 2-32*d* is a conveyor circuit:

Hand   In the Hand position the Stop and start buttons can be used to operate the conveyor, and the amber pilot light will be on.

Off   In the Off position the circuit is inactive and the red conveyor pilot light will be on.

Auto   When the photo switch is actuated based on the presence of an object on the conveyor and the select switch is in the Auto position, the circuit is activated. The photo switch contacts are closed when the conveyor moves the object and turns on the green pilot light. When the object moves out of the sight of the photo switch, contacts open and the conveyor stops.

Caution should be taken when using solid-state devices. Do not short-circuit the output leads, as this can damage the unit. Also always check that the input voltage is within the range of the device.

Inductive Proximity Sensors

Inductive proximity sensors (Fig. 2-33*a, b, c, d*) are designed to operate by generating an electromagnetic field and detecting the eddy current losses generated when ferrous and nonferrous metal target objects enter the field. This means that the drop or change in the magnetic field triggers that sensor.

Shown in Fig. 2-33*e*:

Line 1   Terminal 1, contact of Hand/Off/Auto Selector switch closed in the Hand position as indicated by the XOO, Normally Closed Stop button, terminal 2. Down one line, Normally Open contact of M1 starter. Normally Open Start push button, terminal 3, motor starter with Normally Closed overload contact.

Line 2   Contact of control relay CR1.

Line 3   Output of inductive proximity sensor IPS1, solid-state relay SSR1.

Line 4   Contact of Hand/Off/Auto Selector switch closed in the Auto position as indicated by the OOX, Normally Open contact of the solid-state relay SSR1. Down one line, green pilot light.

Line 5   Contact of Hand/Off/Auto Selector switch closed in the Hand position as indicated by the XOO, amber pilot light.

Line 6   Contact of Hand/Off/Auto Selector switch closed in the Off position as indicated by the OXO.

72   Chapter Two

871TM AC/DC Cable Style
12, 18, 30mm

871TM AC/DC Mini
Quick-Disconnect Style
12, 18, 30mm

871TM AC/DC Micro
Quick-Disconnect Style
12, 18, 30mm

871TM AC/DC EAC Micro
Quick-Disconnect Style
12mm

Figure 2-33a *Courtesy of Rockwell Automation, Inc.*

Inductive Proximity Sensors
871TM 2-Wire AC/DC
Stainless Steel Face/Threaded Stainless Steel Barrel

Dimensions—mm (inches) (continued)
Mini Quick-Disconnect Style

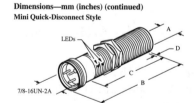

Wiring Diagrams

Normally Open or Normally Closed

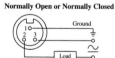

Note 1: No ground pin on 12mm. Attach housing to ground.
Note 2: Load can be switched to pin 3.

Thread Size	Shielded	mm (inches)			
		A	B	C	D
M12 × 1	Y	12.0 (0.47)	85.6 (3.37)	37.8 (1.49)	2.5 (0.10)
	N			31.7 (1.25)	9.4 (0.37)
M18 × 1	Y	18.0 (0.71)	76.6 (3.02)	54.9 (2.16)	2.5 (0.10)
	N			43.1 (1.70)	14.4 (0.56)
M30 × 1.5	Y	30.0 (1.18)	86.4 (3.40)	61.3 (2.41)	2.5 (0.10)
	N			41.6 (1.64)	17.9 (0.70)

Figure 2-33b *Courtesy of Rockwell Automation, Inc.*

Input Devices 73

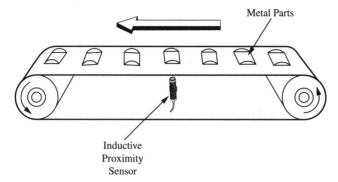

Figure 2-33c Conveyor belt.

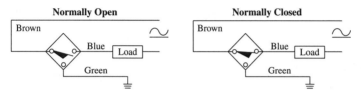

Note 1: No green wire on 12mm and on sensors with PVC cable (–A2).
Attach housing to ground.
Note 2: Load can be switched to brown wire.

Figure 2-33d Wiring diagrams. *Courtesy of Rockwell Automation, Inc. For detailed specifications, see Allen Bradley at www.ab.com.*

Figure 2-33e is a conveyor circuit:

Hand In the Hand position the Stop and Start buttons can be used to operate the pump, and the amber pilot light will be on.

Off In the Off position the circuit is inactive, and the red conveyor pilot light will be on.

Auto When the proximity sensor switch is actuated based on the presence of an object on the conveyor and the select switch is in the Auto position, the circuit is activated. The proximity sensor switch contacts are closed when the conveyor moves the object and turns on the green pilot light. When the object moves out of the sight of the proximity sensor switch, contacts open and the conveyor stops.

Caution should be taken when using solid-state devices. Do not short-circuit the output leads as this can damage the unit. Also always check that the input voltage is within the range of the device.

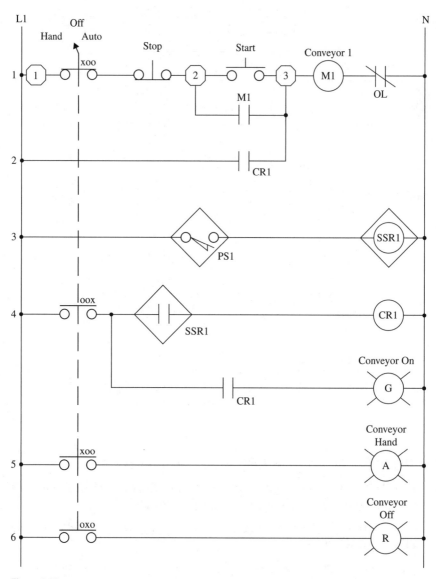

Figure 2-33e

Capacitive Proximity Sensors

Capacitive proximity sensors (Fig. 2-34a, b) are self-contained solid-state devices designed for noncontact sensing of a wide range of materials. Unlike inductive proximity sensors, capacitive proximity sensors can detect nonmetal solids and liquids in addition to standard metal targets. They can even sense the presence of some targets through certain

Figure 2-34a *Courtesy of Rockwell Automation, Inc.*

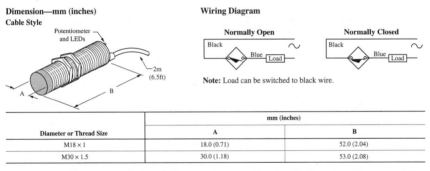

Figure 2-34b *Courtesy of Rockwell Automation, Inc. For detailed specifications, see Allen Bradley at www.ab.com.*

other materials, making them an ideal choice in some applications where inductive proximity and photoelectric sensors cannot be used.

Shown in Fig. 2-34c:

Line 1 Terminal 1, contact of Hand/Off/Auto Selector switch closed in the Hand position as indicated by the XOO, Normally Closed Stop button, terminal 2. Down one line, Normally Open contact of M1 starter. Normally Open Start push button, terminal 3, motor starter with Normally Closed overload contact.

Line 2 Contact of control relay CR1.

Line 3 Output of inductive proximity sensor IPS1, solid-state relay SSR1.

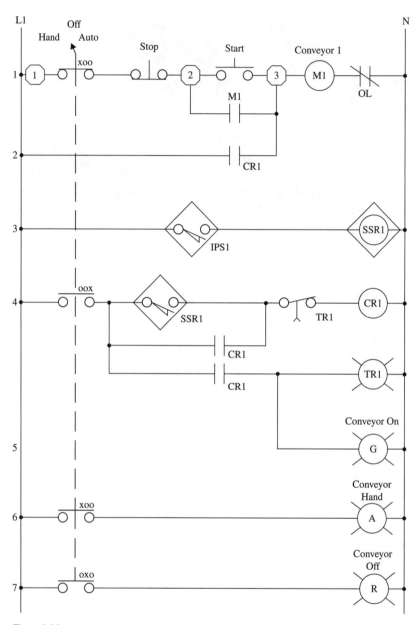

Figure 2-34c

Line 4 Contact of Hand/Off/Auto Selector switch closed in the Auto position as indicated by the OOX. Down one line, open contact of CR1. Down one line, open contact of CR1, time delay relay TR1. Down one line, green pilot light, Normally Open contact of the solid-state relay SSR1, Normally Closed time open contact of TR1, control relay CR1.

Line 6 Contact of Hand/Off/Auto Selector switch closed in the Hand position as indicated by the XOO, amber pilot light.

Line 7 Contact of Hand/Off/Auto Selector switch closed in the Off position as indicated by the OXO.

Figure 2-34c is a conveyor circuit:

Hand In the Hand position the Stop and Start buttons can be used to operate the conveyor, and the amber pilot light will be on.

Off In the Off position the circuit is inactive, and the red conveyor pilot light will be on.

Auto When the proximity sensor switch is actuated based on the presence of an object on the conveyor and the select switch is in the Auto position, the circuit is activated. The proximity sensor switch contacts are closed when the conveyor moves the object and turns on the green pilot light. The conveyor will continue to run for the time that is set by the TR1 relay or when the Selector switch is moved from the Auto position.

Caution should be taken when using solid-state devices. Do not short-circuit the output leads as this can damage the unit. Also always check that the input voltage is within the range of the device.

Limit Switches

There are an unlimited number of limit switches—too many for one book. (See Fig. 2-35a for a few of them.) Limit switches are activated by pressure for an object. They will have at least one contact Normally Open or Closed.

Shown in Fig. 2-35b:

Drawing notes:

LS1—Operator safety cover limit switch. Normally Open held closed. Located at right lower side of the safety cover.

LS2—Box detected limit switch. Located above the conveyor center of platform. Normally Open.

LS3—Left cover. Located bottom right of cover. Normally Open.

LS4—Right cover. Located bottom right of cover. Normally Open.

LS5—Top cover. Located bottom left of cover. Normally Open.

LS6—Bottom cover. Located bottom right of cover. Normally Open.

Specifications	801 General Purpose	802G Gravity Return	802M and 802MC Pre-Wired Factory Sealed	802R Sealed Contact
Description	General purpose limit switch for a wide variety of applications.	Plug-In gravity return switch, designed for conveyor-type operations with small or lightweight objects.	Compact, pre-wired switch is factory sealed to meet the requirements of demanding applications, wet or dry.	Similar in construction to the 802T NonPlug-In. A glass hermetically sealed reed switch is used as the switching element to provide high contact reliability.
Features	Mounting option; Surface	Mounting options: Surface, manifold	The cable entry and wire strands are epoxy sealed to protect against fluids entering or wicking into the switch. Mounting options: surface.	Enclosure: gasketed, transparent plastic cover allows inspection of terminals without removing the cover. Mounting option: Surface
Contact Rating	NEMA A600	NEMA B600	2-circuit: NEMA A600 4-circuit: NEMA B300	NEMA B600
Temperature Rating	$-0°$ to $40°C$ ($32°$ to $104°F$)	$0°$ to $110°C$ ($32°$ to $230°F$)	$0°$ to $80°C$ ($32°$ to $180°F$)	$-29°$ to $121°C$ ($-20°$ to $250°F$)
Actuators	Lever, maintained	Three adjustable rod levers	Lever, maintained, top and side push (with or without rollers).	Lever, low operating force, top and side push (with or without rollers), cat whisker, wobble stick
Enclosure	NEMA Type 1, Type 4 or Type 7 and 9	NEMA Type 1	NEMA Types 1, 4, 4X, 6P and 13;IP67 (IEC529)	NEMA Types 13
Additional Info	See page R5-7	See page R5-11	See page R5-13	See page R5-32

Figure 2-35a Limit switches—quick selection guide. *Courtesy of Rockwell Automation, Inc. For detailed specifications, see Allen Bradley at www.ab.com.*

Line 1 Terminal 1, contact of Hand/Off/Auto Selector switch closed in the Hand position as indicated by the XOO, Normally Closed Stop button, terminal 2. Down one line, Normally Open contact of M1 starter. Normally Open Start push button, terminal 3, motor starter with Normally Closed overload contact.

Line 2 Contact of control relay CR1.

Line 3 LS1, control relay CR2.

Line 4 Contact of Hand/Off/Auto Selector switch closed in the Auto position as indicated by the OOX. Down one line, open contact of CR1. Down one line, open contact of CR1, time delay relay TR1. Down one line, green pilot light, Normally Open contact of limit switch LS2, Normally Closed time open contact of TR1, control relay CR1.

Line 6 Contact of Hand/Off/Auto Selector switch closed in the Hand position as indicated by the XOO, amber pilot light.

Line 7 Contact of Hand/Off/Auto Selector switch closed in the Off position as indicated by the OXO.

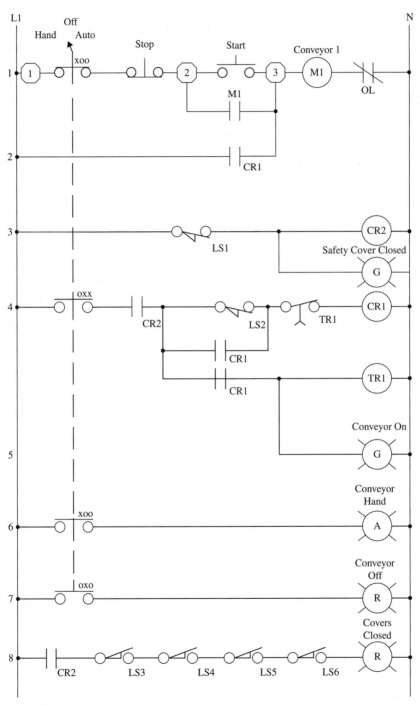

Figure 2-35*b*

Line 8 Contact of CR2 Normally Open, Normally Open contact of limit switch LS3, Normally Open contact of limit switch LS4, Normally Open contact of limit switch LS5, Normally Open contact of limit switch LS6, red pilot light cover closed.

Figure 2-35*b* is an example of a conveyor circuit:

Hand In the Hand position the Stop and Start buttons can be used to operate the conveyor, and the amber pilot light will be on.

Off In the Off position the circuit is inactive, and the red conveyor pilot light will be on.

Auto The operator safety cover must be closed. When limit switch LS2 is actuated based on the presence of a box on the conveyor and the select switch is in the Auto position, the circuit is activated. The limit switch LS2 contacts are closed; the conveyor moves the object and turns on the green pilot light. The conveyor will continue to run for the time that is set by the TR1 relay or the Selector switch is moved from the Auto position.

Chapter

3

Output Devices

I define an *output device* as a device that is the final element in a control circuit. It controls or annunciates an electrical circuit. As you can imagine, there are an unlimited number of output devices. I will cover the most common ones. There are many manufacturers of these devices; for the most part I will use Rockwell Automation's Allen Bradley devices. I am most familiar with them and they are widely used in the field. The information contained here is not meant to replace the manufacturer's specifications; for detailed data please see the manufacturer's specifications.

Relays

Relays come in all shapes and sizes, from nanorelays to very large relays. Basically a relay is an electrical switch that opens or closes a control circuit. An electromagnetic relay is operated by an electromagnet that opens or closes electrical contacts when power is applied to the electromagnet. The position of the contacts changes by spring, gravity action, or permanent magnet when the electromagnet is deenergized. Most relays are used in control applications. The contacts are generally in the range of 0 to 35 A. All relays have at least one contact. Relays used in control systems will typically have from one to eight contacts. The contacts can be Normally Open or Normally Closed. Some relays have contacts added or can be changed in the field from open to closed or from closed to open. There are timing relays that are used to control events by time, and there are latching relays that remember their last state. Relays are used to give multiple contacts to an input or pilot device and can extend the control circuit to many different circuits with different voltages or requirements. As you will see in Fig. 3-1, one relay CR1 can control several devices, and the output contacts can control any voltage and current within the range of the relay's contact specifications.

Shown in Fig. 3-1:

Line 1 Normally Open flow FS1 and control relay CR1. Normally Open flow switch closes and energizes CR1 when the liquid rises to a preset level.

Line 2 Normally Open contact of relay CR1, pump motor starter M1, Normally Closed overload contact of motor starter M1. When CR1 is energized, the contact closes and the pump motor starter is energized. The pump runs until the liquid level drops to a preset level.

Line 3 Normally Open contact of relay CR1, green pilot light. When CR1 is energized, the contact closes and the green pilot light is illuminated until the liquid level drops to a preset level.

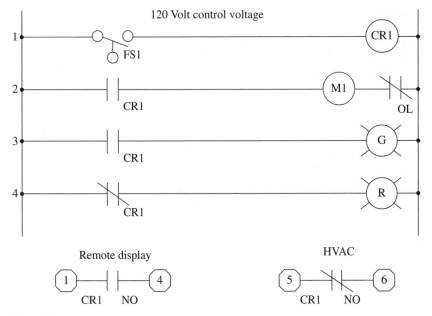

Figure 3-1

Line 4 Normally Closed contact of relay CR1, red pilot light. When CR1 is energized, the contact opens and the red pilot light is turned off until the liquid levels drop to a preset level.

Output contacts

1. *Normally Open contact terminals 3 and 4.* When the control relay CR1 is energized, the contact will close and illuminate a remote display until the liquid level drops to a preset level. This control circuit is 24 V DC.

2. *Normally Closed contact terminals 5 and 6.* When the control relay CR1 is energized, the contact will open and the HVAC control circuit will be opened until the liquid level drops to a preset level. This control circuit is 48 V AC.

Open Frame Relay

Figure 3-2 is an open frame relay. These relays are used in a dust-proof enclosure or in an environment that has no dust in the air. This relay is available in single-pole single-throw double-make with one Normally Open contact (SPST-NO-DM), single-pole double-throw with

84   Chapter Three

Figure 3-2 700-HG. *Courtesy of Rockwell Automation, Inc. For detailed specifications, see Allen Bradley at www.ab.com.*

one Normally Open contact and one Normally Closed contact (SPDT), double-pole single-throw Normally Open (DPST-NO), and double-pole double-throw (DPDT). Coil voltage is from 24 to 480 V AC. Contacts are rated at 40 A. See Fig. 3-2*a*, *b*, *c,* and *d*.

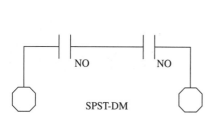

Figure 3-2*a*

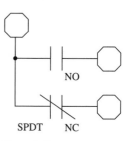

Figure 3-2*b*

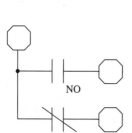

Figure 3-2*c*

Figure 3-2*d*

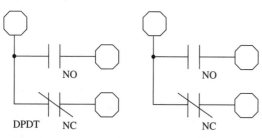

Figure 3-2*e*

Output Devices   85

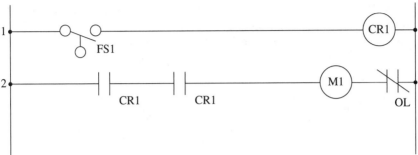

Figure 3-2f

Shown in Fig. 3-2f:

Line 1 Float switch FS1 and control relay CR1. When level reaches a preset amount, the float switch closes and energizes control relay CR1. CR1 remains energized until level drops below a preset amount.

Line 2 Normally Open contact CR1, Normally Open contact CR1, motor starter M1. When CR1 is energized, CR1 contacts close and motor starter M1 is energized; the pump runs until CR1 is deenergized.

Heavy-Duty Industrial Relays

Heavy-duty industrial relays (Fig. 3-3) are electromechanical relays. The coil voltage can be 5 to 600 V AC. These relays are used in control circuits. They have typically from one to eight contacts, which can be any combination of Normally Open and Normally Closed contacts.

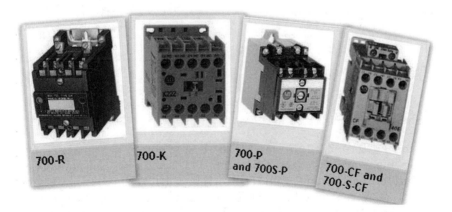

Figure 3-3 *Courtesy of Rockwell Automation, Inc. For detailed specifications, see Allen Bradley at www.ab.com.*

Heavy-duty industrial relays can have field-changable contacts. They are available with sealed, reed, or standard contacts, and these contacts can be mixed to meet the field requirements. Contacts are rated from 5 to 35 A. Timer decks are available for some types of these relays. The timer decks extend the function to allow time control.

Figure 3-3*a*   Figure 3-3*b*

In the circuit in Fig. 3-3*c*, one three-wire start/stop station is used to control three motors using a control relay.

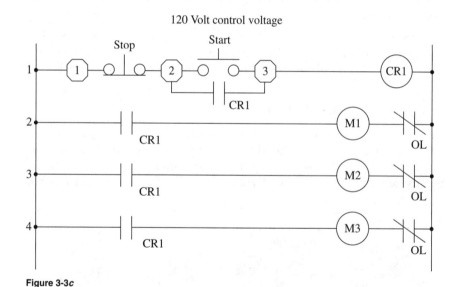

Figure 3-3*c*

Shown in Fig. 3-3*c*:

Line 1 Terminal 1, Normally Closed Stop button, terminal 2. Down one line, Normally Open CR1 contact. Up one line, Normally Open Start button, terminal 3, and control relay CR1. When the Start button is pressed, control relay CR1 is energized, CR1 contacts are closed, and the relay is electrically latched. The control relay CR1 will remain energized until the Stop button is pressed.

Line 2 Normally Open contact of control relay CR1, motor starter M1, and overload contact of motor starter M1. When control relay CR1 is energized, then contact CR1 will close, the starter will be energized, and the motor will run.

Line 3 Normally Open contact of control relay CR1, motor starter M2, and overload contact of motor starter M2. When control relay CR1 is energized, then contact CR1 will close, the starter will be energized, and the motor will run.

Line 4 Normally Open contact of control relay CR1, motor starter M3, and overload contact of motor starter M3. When control relay CR1 is energized, then contact CR1 will close, the starter will be energized, and the motor will run.

Sealed Relay (Ice Cube)

In Fig. 3-4, we see the sealed relay, also known as an ice cube relay. These relays can be socket-based or soldered. They can be base-mounted or flange-mounted. Contacts will vary with the relay selected. This relay has single-pole single-throw double-make with one Normally Open contact (SPST-DM-NO), double-pole double-throw (DPDT), and three-pole double-pole double-throw (3PDT), and is available from 24 to 120 V AC and 6 to 24 V DC. Contacts are rated at 20, 25, and 30 A. See Fig. 3-4*a*, *b*, and *c*.

Figure 3-4 *Courtesy of Rockwell Automation, Inc. For detailed specifications, see Allen Bradley at www.ab.com.*

Figure 3-4a

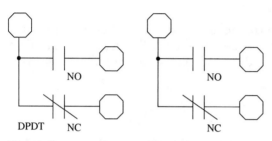

Figure 3-4b

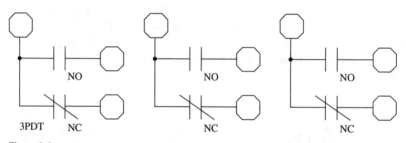

Figure 3-4c

In Fig. 3-4d I use an SPST-DM contact.

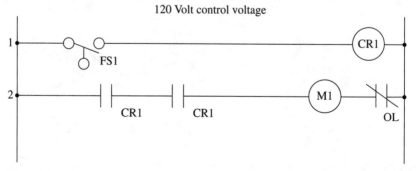

Figure 3-4d

Shown in Fig. 3-4*d*:

Line 1 Float switch FS1 and control relay CR1. When the level reaches a preset amount, the float switch closes and energizes control relay CR1. CR1 remains energized until the level drops below a preset amount.

Line 2 Normally Open contact CR1, Normally Open contact CR1 motor starter M1. When CR1 is energized, then CR1 contacts close, motor starter M1 is energized, and the pump runs until CR1 is deenergized.

Figure 3-4*e*, *f*, *g*, and *h* comprises ice cube relay pin outs. These are standard pin outs, but despite all so-called standards, they can change from manufacturer to manufacturer. Most ice cube relays will have the pin outs stamped on the top of the relay. It is a good idea to check the pin outs before connecting the relay. As you can see in the 8-pin relay base in Fig. 3-4*e*, the

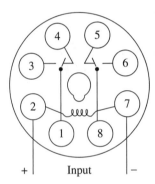

Figure 3-4e *Courtesy of Rockwell Automation, Inc.*

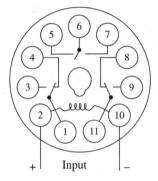

Figure 3-4f *Courtesy of Rockwell Automation, Inc.*

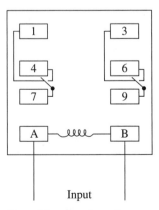

Figure 3-4g *Courtesy of Rockwell Automation, Inc.*

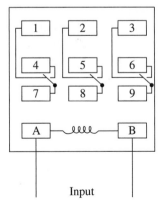

Figure 3-4h *Courtesy of Rockwell Automation, Inc.*

coil is pins 2 and 7. On the 11-pin relay in Fig. 3-4f, the coil is pins 2 and 10. The flat blade relays are a little more consistent, as you can see in Fig. 3-4g and h. It is always a good idea to check it out before you hook it up.

Latching Relays

Latching relays are available in many configurations—heavy-duty industrial relays, high-current contact relays, and sealed relays also known as ice cube relays. Figure 3-5 shows a mechanical latching relay. The contacts remain in their last state until reset. These relays can have

Figure 3-5 *Courtesy of Rockwell Automation, Inc. For detailed specifications, see Allen Bradley at www.ab.com.*

one AC coil or two DC coils. They can be panel-mounted, socket-based, or soldered. This relay has single-pole double-throw with one Normally Open contact and one Normally Closed contact (SPDT), double-pole double-throw (DPDT) with one coil, and double-pole double-throw (DPDT) with dual coils; this relay is available as single-coil AC or dual-coil DC. See Fig. 3-5a and b.

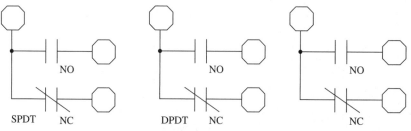

Figure 3-5a   **Figure 3-5b**

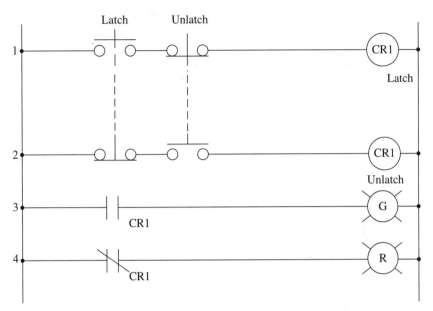

Figure 3-5c

Shown in Fig. 3-5c:

Line 1 Normally Open latch push button, Normally Closed unlatch push button, and CR1 latch coil.

Line 2 Normally Closed latch push button, Normally Open unlatch push button, and CR1 unlatch coil.

Line 3 Normally Open CR1 contact, green pilot light.

Line 4 Normally Closed CR1 contact, red pilot light.

When the latch push button is pressed, it closes the contact in line 1 and latches the relay and opens the contact in line 2 so that the unlatch and latch coils are not energized together. The relay closes the contact in line 3, and the green pilot light is illuminated. The relay opens the contact in line 4 and turns off the red pilot light. The relay will remain latched, even if power is lost, until the unlatch push button is pressed.

When the unlatch push button is pressed, it opens the contact in line 1, so that the unlatch and latch coils are not energized together, and closes the contact in line 2 and unlatches the relay. The relay opens the contact in line 3, and the green pilot light turns off. The relay closes the contact in line 4, and the red pilot light is illuminated. The relay will remain unlatched until the latch push button is pressed.

Timing Relays

Timing relays are available in many configurations. They can have pneumatic or solid-state timers. Pneumatic timers will continue to operate when the power is lost. Timers can have instantaneous contacts. Instantaneous contacts are operated directly by the relay coil and are not time-controlled. This combines the function of a standard relay and a timer relay. Solid-state timers are more accurate and have a greater range of control times. Timing relays can be time off delay (time delay

Figure 3-6 700-FS. *Courtesy of Rockwell Automation, Inc. For detailed specifications, see Allen Bradley at www.ab.com.*

starts when the relay is deenergized) or time on delay (time delay starts when the relay is energized). The relay shown in Fig. 3-6 allows both configurations. It has single-pole double-throw with one Normally Open contact and one Normally Closed contact (SPDT), Normally Open contact NO. See Fig. 3-6a to d.

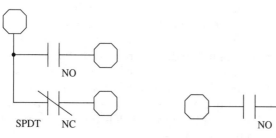

Figure 3-6a

Figure 3-6b

Output Devices   93

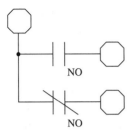

Figure 3-6c

Figure 3-6d

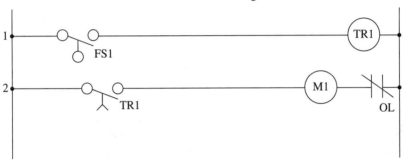

Figure 3-6e

Shown in Fig. 3-6e:

Line 1 Float switch FS1, timer relay TR1. When the level reaches a preset amount, the float switch closes and starts the time cycle.

Line 2 Normally Open time-close contact, pump motor starter M1, Normally Closed overload. When the TR1 is energized, it starts the

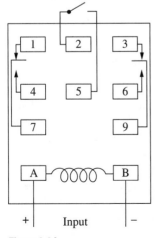

Figure 3-6f

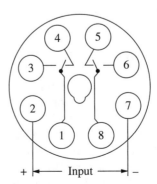

Figure 3-6g

time preset cycle; when the time cycle is complete, the contact TR1 will close, the motor starter will be energized, and the pump will start. This allows the level to fluctuate without the pump running intermittently.

Figure 3-6*f* is the pin out for the 700-HS, and Fig. 3-6*g* is the pin out for the 700-HV; both are ice cube–style timer relays. These can change from manufacturer to manufacturer. The pin outs are usually printed on the relay. The pin out should be read before connecting the timer.

Solid-State Relays

The solid-state relays shown in Fig. 3-7 have several advantages over standard relays. Solid-state relays have no moving parts and are impervious to shock and vibrations. They are sealed from moisture and dirt.

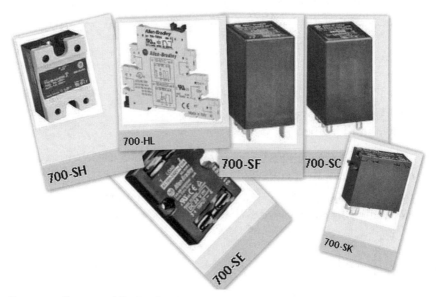

Figure 3-7 *Courtesy of Rockwell Automation, Inc. For detailed specifications, see Allen Bradley at www.ab.com.*

The major advantage of solid-state relays is that the input and output circuits are completely isolated. This prevents spikes and noise in the control circuit from being transmitted to the controlled circuit. Solid-state relays can be used to control AC or DC circuits.

The solid-state relay typically uses a reed relay or photoelectric device to isolate the circuits. The reed relay is operated by an electromagnetic coil to close a sealed reed contact. In a photoelectric isolated solid-state relay, an LED is used to operate a photosensitive switch in an AC circuit; this would be a TRIAC. In a DC circuit it would be a transistor.

Solid-state relays are controlled by low voltage and current. This allows for the use of smaller conductors.

Solid-state relays are sensitive to heat and generate a lot of heat. They should be mounted on a heat sink as shown in Fig. 3-7e and should have good ventilation.

Solid-state relays typically have one Normally Open contact (SPST) or one Normally Open and one Normally Closed contact (SPDT) (reed-type only).

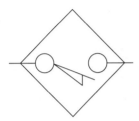

Figure 3-7a

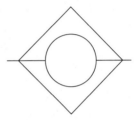

Figure 3-7b

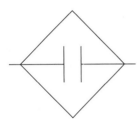

Figure 3-7c

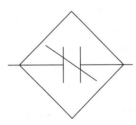

Figure 3-7d

Figure 3-7e *Courtesy of Rockwell Automation, Inc. For detailed specifications, see Allen Bradley at www.ab.com.*

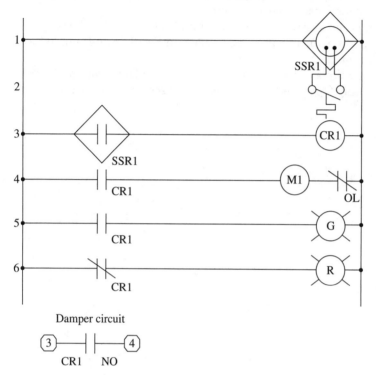

Figure 3-7f

Figure 3-7f illustrates how to amplify the solid-state relay to control many circuits.

Line 1 Supply continuous power to solid-state relay SSR1.

Line 2 Temperature switch closes at a preset temperature and triggers SSR1.

Line 3 Normally Open contact of solid-state relay SSR1 and control relay CR1. When SSR1 is triggered closed, CR1 will be energized.

Line 4 Normally Open contact of CR1, motor starter M1, and Normally Closed overload contact of M1. When CR1 is energized, the contact will close and the starter M1 will be energized, starting the blower motor.

Line 5 Normally Open contact of CR1, green pilot light. When CR1 is energized, the green pilot light will be illuminated.

Line 6 Normally Closed contact of CR1, red pilot light. When CR1 is energized, the contact CR1 will open and the red pilot light will be off.

Output Normally Open contact of CR1. When CR1 is energized, the contact will close, completing the damper circuit

Solid-State Timer Relays

The solid-state timers in Fig. 3-8 can have standard electromechanical coils or solid-state coils. Solid-state timers use a potentiometer to set or adjust the timing cycle. The potentiometer can be internal or external.

Figure 3-8 *Courtesy of Rockwell Automation, Inc. For detailed specifications, see Allen Bradley at www.ab.com.*

Care should be taken not to apply voltage to the potentiometer circuit, as this will damage the timer. Timers with solid-state inputs typically use a Normally Open contact to initiate the timing cycle. Care should be taken not to apply voltage to the initiating circuit, as this will damage the timer. Solid-state timers must have power applied to operate the timer. If the power fails, the timing cycle will terminate. Length of the timing cycle can be from a fraction of a second to hours.

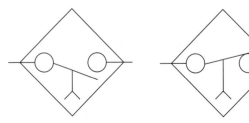

Figure 3-8a   **Figure 3-8b**

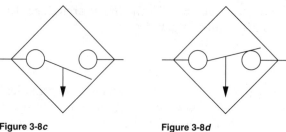

Figure 3-8c   Figure 3-8d

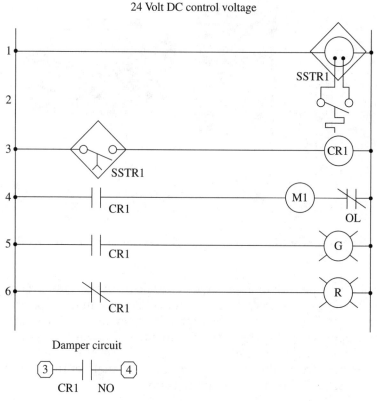

Figure 3-8e

Figure 3-8e illustrates how to amplify the solid-state timer relay to control many circuits.

Line 1   Supply continuous power to solid-state timer relay SSTR1.

Line 2   Temperature switch closes at a preset temperature and triggers SSTR1.

Line 3 Normally Open time-close contact of solid-state relay SSTR1 and control relay CR1. When SSTR1 is triggered, preset time cycle starts; when complete contact is closed, CR1 will be energized.

Line 4 Normally Open contact of CR1, motor starter M1, and Normally Closed overload contact of M1. When CR1 is energized, the contact will close and the start M1 will be energized, starting the blower motor.

Line 5 Normally Open contact of CR1, green pilot light. When CR1 is energized, the green pilot light will be illuminated.

Line 6 Normally Closed contact of CR1, red pilot light. When CR1 is energized, the contact CR1 will open and the red pilot light will be off.

Output Normally Open contact of CR1. When CR1 is energized, the contact will close, completing the damper circuit.

Shunt Trip Breaker

A shunt trip breaker such as those in Fig. 3-9 has an electromechanical relay that allows the breaker to be tripped remotely by a low-voltage control circuit. The coil can be any voltage offered by the manufacturer and

Figure 3-9 *Courtesy of Rockwell Automation, Inc. For detailed specifications, see Allen Bradley at www.ab.com.*

is usually in the range of 12 to 600 V AC/DC. The shunt trip relay allows systems such as the fire alarm or emergency power off (EPO) system to trip the circuit breaker in case of an alarm event. It is important that the shunt breaker be connected and tested. NEC Section 645.10 states that an IT equipment room must have remote disconnect of all power in the room.

Figure 3-9a   Figure 3-9b

In Fig. 3-9c there are two ways to trip the shunt trip: Normally Open circuit and Normally Closed circuit.

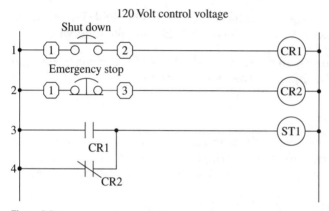

Figure 3-9c

Shown in Fig. 3-9c:

Line 1 Terminal 1, Normally Open Mushroom push button, terminal 2, and control relay CR1. When the Shutdown push button is pressed, CR1 is energized.

Line 2 Terminal 1, Normally Closed Mushroom push button, terminal 3, and control relay CR2. When the Shutdown push button is pressed, CR2 is deenergized.

Line 3 Normally Open contact of CR1 and shunt trip coil ST1. When CR1 is energized, the CR1 contact will close and the breaker will trip.

Line 4 Normally Closed contact of CR2. When CR2 is deenergized, the CR2 contact will close and the breaker will trip.

Three-Phase Motor Starter

Motor starters such as that in Fig. 3-10 are electromechanical relays that allow the control of high voltage and current with the line or low-voltage control circuit.

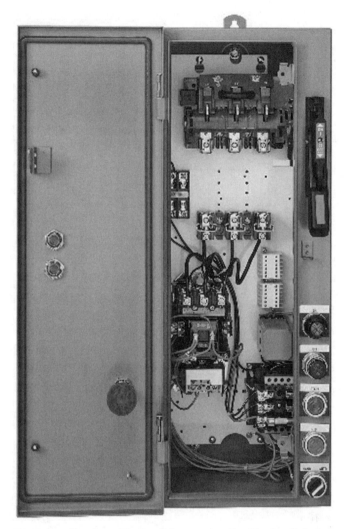

Figure 3-10 *Courtesy of Rockwell Automation, Inc. For detailed specifications, see Allen Bradley at www.ab.com.*

In the motor starter circuit in Fig. 3-10a, L1, L2, and L3 are connected to a circuit breaker to protect the circuit. The control transformer is used to reduce the line voltage to a 120-V control circuit. The control transformer is connected to line 1 and line 2. The control circuit is protected but fuses FU. The fuses are connected to the transformer terminal H1 and H4. The X1 terminal of the control transformer is connected to fuse FU to protect the control circuit devices and wires. L1 is the high, or hot, line of the control circuit. The X2 terminal is grounded and is the neutral side of the control circuit.

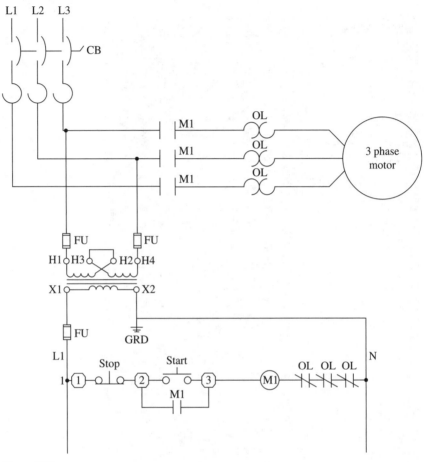

Figure 3-10a

Shown in Fig. 3-10a:

Line 1 Terminal 1, Normally Closed Stop push button, terminal 2. Down one line, Normally Open contact of motor starter coil M1. Up one line, Normally Open Start push button, terminal 3, motor starter coil M1, and the Normally Closed overload contacts of M1. When the Start button is pushed, the M1 coil is energized, the M1 contacts close, and the motor starts. The M1 contact closes and electrically latches the circuit. The motor will continue to run until the Stop push button is pressed or power to the control circuit is shut off.

Chapter

4

Monitoring Systems

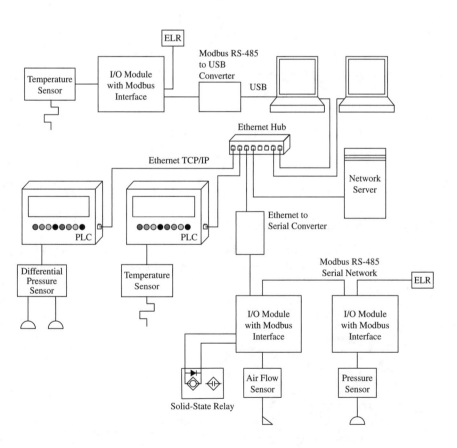

103

I define *monitoring system* as a device or devices that monitor conditions, alarms, and events and report them to a data collection computer system or programmable logic controller (PLC).

Figure 4-1 is a computer-based monitoring system. This type of system is used to monitor and collect data from the process or environments, and it can have some control functions, which are usually limited to the control of a few control I/O. When more complex control is required, then a PLC is used to control and collect data. The network shown has the most commonly used network protocols. The types of networks that are most used in the field for monitoring are the serial daisy-chained network RS-232, serial multipoint network RS-485, USB interface, and Ethernet topology. I will look at each, and we will see how they are used. Each one is explored in the following pages.

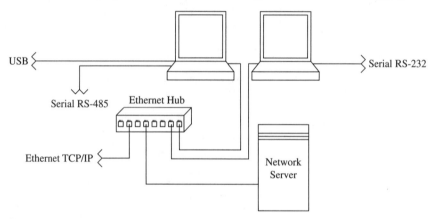

Figure 4-1

Programmable Logic Controllers

Programmable logic controllers such as the Allen Bradley Micrologix 1100 in Fig. 4-2 are used to collect and control systems with a couple

Figure 4-2 *Courtesy of Rockwell Automation, Inc. For detailed specifications, see Allen Bradley at www.ab.com.*

of inputs to 80 digital and analog inputs and outputs. For the purposes of this book we will look at the data collection of a few I/O. One of the main advantages of using a PLC for data collection is that it can communicate the information on multiple networks. The Micrologix 1100 can use 10/100 Mbps Ethernet/IP and RS-3232 and RS-485, which makes it very flexible as a monitoring/data collection system.

Monitoring and Control Modules

Monitoring and control modules are devices that allow the monitoring and control of processes and environmental conditions. Modules use RS-232 and RS-485 and Modbus protocols to communicate with monitoring computer systems. A lot of equipment today, such as computer room air conditioning (CRAC) units and kilowatt meters, have built-in modules that support Modbus on RS-485 communication. When equipment that needs to be monitored does not have an internal Modbus module, external modules such as the DGH Corporation modules in Fig. 4-3 can be used. Modules and devices like those offered by DGH Corporation also allow for the conversion from one network topology to another. For a complete list of modules and converters offered by DGH Corporation or for specifications, visit their website at www.dghcorp.com.

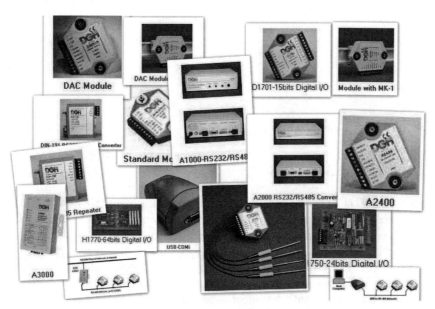

Figure 4-3 *Courtesy of DGH Corporation. See detailed specifications at www.dghcorp.com.*

Figure 4-4 is a very busy diagram and uses many network protocols and interfaces. In the next pages we will take the network apart and look at each type of system.

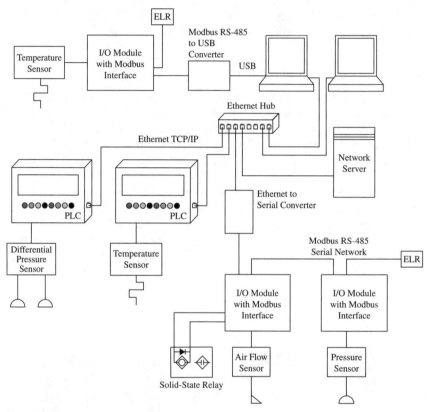

Figure 4-4

RS-232 Introduction and Specifications

RS-232, the best-known serial interface, is implemented on almost all computers available today.

RS-232 is an interface most often used to connect one data terminal equipment (DTE) to one data communication equipment (DCE) at a maximum speed of 20 kbps with a maximum cable length of 50 ft or the cable length equal to a capacitance of 2500 pF. [1 picofarad (pF) = 1 trillionth of a farad (F).] This latter rule is often forgotten. This means that using a cable with low capacitance allows you to span longer distances without going beyond the limitations of the standard. If, for example, UTP CAT-5 cable is used with a typical capacitance of 17 pF/ft, the maximum allowed cable length is 2500 pF/17 pF = 147 ft. This is sufficient where computer equipment is connected using a modem or short cables. Using twisted-pair shielded cable will reduce or eliminate noise and communication errors. Line noise is the chief cause of communication errors (see Fig. 4-6a).

Interface communication as defined in the RS-232 standard is an asynchronous serial communication method. The term *serial* means that the information is sent 1 bit at a time. The term *asynchronous* tells us that the information is not sent in predefined time slots. Data transfer can start at any time, and it is the task of the receiver to detect when a message starts and ends.

RS-232 DB9 Pinout

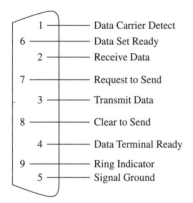

Figure 4-4a

RS-232 DB25 Pinout

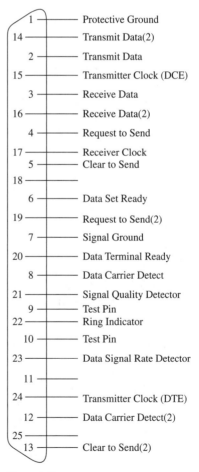

Figure 4-4b

Devices in a daisy-chain RS-232 network are looped in series. Please refer to Fig. 4-4e.

The original pin out for RS-232 was developed for a 25-pin sub D connector. Since the introduction of the smaller serial port on the IBM-AT, 9-pin RS-232 connectors are commonly used. Most of the computers now have USB ports; if equipped with a serial port, they will have the DB9 serial port version.

The following table compares DB9 to DB25:

DB9	DB25	Function
1	8	Data Carrier Detect
2	3	Receive Data
3	2	Transmit Data
4	20	Data Terminal Ready
5	7	Signal Ground
6	6	Data Set Ready
7	4	Request to Send
8	5	Clear to Send
9	22	Ring Indicator

Figure 4-4c shows typical RS-232 connections. In this example there is a modem connected to a single computer. This is the most used configuration of the RS-232 interface.

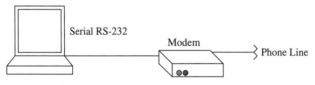

Figure 4-4c

Figure 4-4d is an example of a daisy-chained RS-232 network. RS-232 multidrop networks can have 124 devices of DGH Corporation modules in the network.

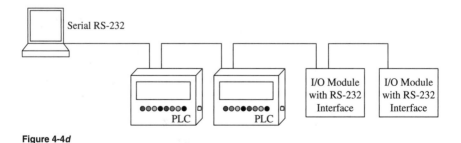

Figure 4-4d

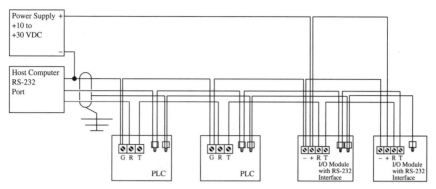

Figure 4-4e

Figure 4-4e is a wiring diagram of an RS-232 daisy-chain network. Shielded cable is used to reduce or eliminate noise (see Fig. 4-6a). When terminating the wire, make the exposed (stripped) end of the wire as short as possible, and keep it away from power conductors to minimize the introduction of noise. This diagram is based on the specification for the DGH Corporation modules. Check the job specification to ensure proper wiring of the modules.

1. Host computer with RS-232 serial port. This will usually be a 9-pin connector; see Fig. 4-4a for pin out.

2. Green pin 5 signal ground, on a 9-pin connector. This is not the shield earth or frame ground. Make sure that the connection is electrically and mechanically secure. If the wire should come loose, the entire network will be lost.

3. Black pin 2 receive data, on a 9-pin connector, is looped through the devices. I like to terminate the wire with a crimp-type wire nut to ensure that the connection will not vibrate loose. On the last device in the chain it is terminated on the T (transmit data). Make sure that the connection is electrically and mechanically secure. If the wire should come loose, the entire network will be lost.

4. Red pin 3 transmit data, on a 9-pin connector, is connected to the R (receive data) in the first device. Then it is connected to T (transmit data) to loop to the next device. In the last device it is terminated on the R terminal. Make sure that the connection is electrically and mechanically secure. If the wire should come loose, the entire network will be lost.

5. Blue is the shield that is grounded in the panel to the earth or frame ground. The shield is only grounded at one end in the panel. This is to prevent a ground loop and eliminate noise. I like to use

heat shrink tubing to insulate the ground. If this is not possible, then tape the shield ground so that it does not short or touch the ground. Use the good tape—not the 12 rolls for a dollar stuff. I like to terminate the wire with a crimp-type wire nut to ensure that the connection will not vibrate loose.

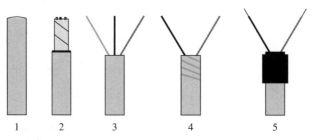

1   2   3   4   5

Figure 4-4f

Figure 4-4f is a method for terminating a shielded twisted-pair cable.

1. Shielded twisted-pair cable.
2. Strip the cable, keeping the stripped wire as short as possible, being careful not to nick the wires and expose the shield. Keep the stripped wire away from power conductors to minimize noise.
3. Remove the shield and separate the wires. In Fig. 4-4f the light gray wire is the shield, the black wire is the R (receive data), and the dark gray wire is the T (transmit data).
4. Twist the shield back around the wire to keep it from grounding or shorting.
5. Insulate the shield. I like to use heat shrink tubing. If that is not possible, then wrap tape around the shield, insulating the shield. If you use tape, then use the good tape; it is important that this does not come loose with heat or humidity.

Figure 4-4g is a method for terminating a shielded twisted-pair cable for an RS-232 loop network.

1. Shielded twisted-pair cables.
2. Strip each of the cables, keeping the stripped wire as short as possible, being careful not to nick the wires and expose the shield. Keep the stripped wire away from power conductors to minimize noise.
3. Remove the shield and separate the wires. In Fig. 4-4g the light gray wire is the shield, the black wire is the R (receive data), and the dark gray wire is the T (transmit data).

Monitoring Systems   111

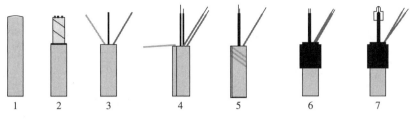

Figure 4-4g

4. Join the cables and twist the shield together.
5. Twist the shields back around the wire to keep it from grounding or shorting.
6. Insulate the shield. I like to use heat shrink tubing. If that is not possible, then wrap tape around the shield, insulating the shield. If you use tape, then use good tape; it is important that this does not come loose with heat or humidity.
7. Twist the stripped ends of the black wires [R (receive data)] together and connect using a crimp-type wire nut.

Figure 4-5 shows an Allen Bradley 837E temperature sensor with 4- to 20-mA output; an Allen Bradley MicroLogix 1100 PLC with RS-232,

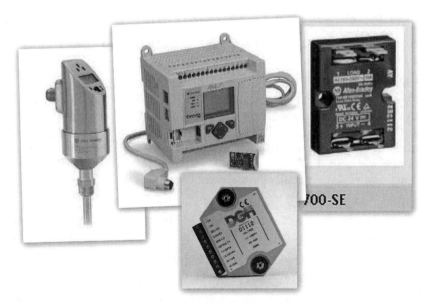

Figure 4-5 *Courtesy of Rockwell Automation, Inc.; for detailed specifications, see Allen Bradley at www.ab.com. Courtesy of DGH Corporation; see detailed specifications at www.dghcorp.com.*

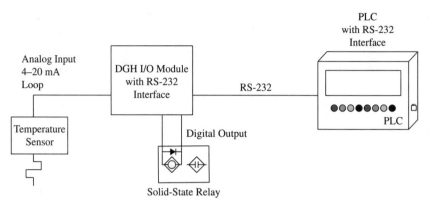

Figure 4-5a

RS-485, and Ethernet; an Allen Bradley 700SE solid-state relay; and a DGH Corporation D1000 RS-232 I/O module with analog input and digital output.

Shown in Fig. 4-5b:

Line 1   Power for SSR1.

Line 2   Power for DGH1.

Line 3   Analog input of 4 to 20 mA to DGH1.

Line 4   Power to temperature sensor. Output from sensor of 4 to 20 mA to input of DGH1.

Line 5   Normally Open contact SSR1, CR1 coil. When the temperature sensor is in the preset range, the DGH1 module outputs a signal to the digital output, which energizes the SSR1, closes the contact, and holds it closed until the temperature is outside of the preset range. This energizes control relay CR1. DGH1 will communicate the temperature to the PLC over the RS-232 network, and the PLC can react and run the required program.

Line 6   Normally Open contact CR1, coil motor starter M1, Normally Closed overload contact of M1. When CR1 is energized, the motor will start the blower motor.

Line 7   Normally Open contact CR1, green pilot light. When CR1 is energized, the green pilot light will be illuminated.

Line 8   Normally Closed contact CR1, red pilot light. The red pilot light will be illuminated until CR1 is energized.

Monitoring Systems 113

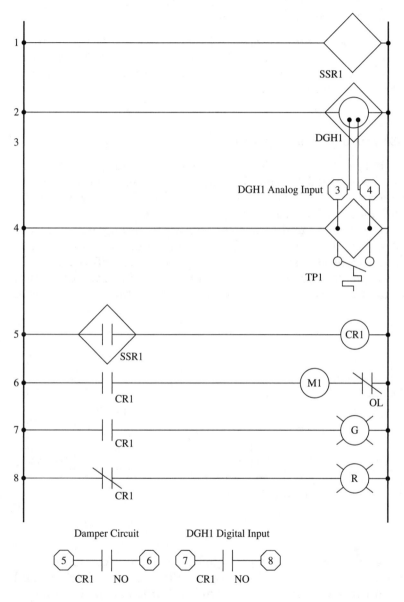

Figure 4-5b

Output contacts

Normally Open contact terminals 5 and 6. When CR1 is energized, the contact will close and open the dampers.

Normally Open contact terminals 7 and 8. When CR1 is energized, the contact will close and close the digital input to DGH1, which will communicate to the PLC over the RS-232 network, and the PLC can react and run the required program.

RS-485 Introduction and Specifications

RS-485 is the most versatile communication standard in the standard series defined by the EIA. RS-485 connects data terminal equipment (DTE)/remote terminal unit (RTU) directly without the need for modems and connects several DTEs/RTUs in a network structure. It has the ability to communicate over long distances and virtually unlimited nodes with the use of repeaters. It has the ability to communicate at faster speeds. This is why RS-485 is currently the most widely used communication interface in data acquisition and control applications where multiple nodes communicate with one another.

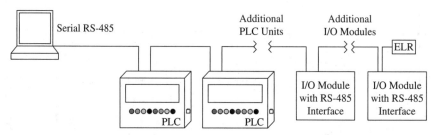

Figure 4-6 Typical RS-485 multipoint network.

The signal is measured from signal + to signal − communication lines; this differential signal method results in longer distances and higher bit rates. One of the most significant problems with RS-232 is the lack of immunity for noise on the communication lines. The transmitter and receiver compare the voltages of the data and handshake lines with signal ground common zero line. Changes in the ground level can have disastrous effects. Therefore the level of the RS-232 interface is relatively high at ±3 V. Noise is easily induced and limits both the maximum distance and communication speed. With RS-485, on the other hand, there is no such thing as a common zero as a signal reference. A difference of several volts between the ground level of the RS-485 transmitter and receiver does not cause any problems. The RS-485 signals are floating, and each signal is transmitted over a signal + line and a signal − line. The RS-485 receiver compares the *voltage difference* between both lines, instead of the absolute *voltage level* on a signal line. This works well and prevents the existence of ground loops, a common source of communication problems. The best results are achieved if the signal + and signal − communication lines are twisted and separately shielded. Figure 4-6a explains this concept.

Figure 4-6a shows the effect of a magnetic field, which can generate noise on the communication lines.

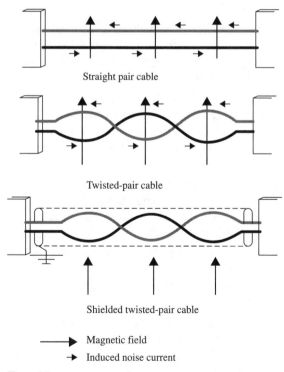

Figure 4-6a

Straight pair cable allows the magnetic field to induce noise spikes on the communication lines.

Twisted-pair cable has induced noise in opposing directions, thereby canceling the noise.

Shielded twisted-pair cable is virtually noise-free. The shield blocks the magnetic field from inducing noise in the communication lines. The only way that noise can affect the communication lines is at the stripped end. This is why the stripped end should be as short as possible.

Figure 4-6b is a wiring diagram of an RS-485 multipoint network. Shielded cable is used to reduce or eliminate noise. When terminating the wire, make the exposed (stripped) end of the wire as short as possible, and keep it away from power conductors to minimize the introduction of noise. This diagram is based on the specification for the DGH Corporation modules. Check the job specification to ensure proper wiring of the modules.

1. Host computer with RS-485 serial port. This will usually be a 9-pin connector; see Fig. 4-4a for pin out.
2. Green signal − is looped through the devices and connected on the signal − terminal (G). This is not connected to earth or frame ground. Make sure that the connection is electrically and mechanically secure. If the wire should come loose, the network from this point forward will be lost.
3. Orange signal + is looped through the devices and connected on the signal + terminal (Y). Make sure that the connection is electrically and mechanically secure. If the wire should come loose, the network from this point forward will be lost.
4. Red power positive + is looped through the devices and connected on the signal + terminal (R). Make sure that the connection is electrically and mechanically secure. If the wire should come loose, the network from this point forward will be lost.
5. Black power negative − is looped through the devices and connected on the signal + terminal (B). Make sure that the connection is electrically and mechanically secure. If the wire should come loose, the network from this point forward will be lost.
6. Blue is the shield that is grounded in the panel to the earth or frame ground. The shield is only grounded at one end in the panel. This is to prevent a ground loop and to eliminate noise. I like to use heat shrink tubing to insulate the ground. If this is not possible, then tape the shield ground so that it does not short or touch the ground. Use the

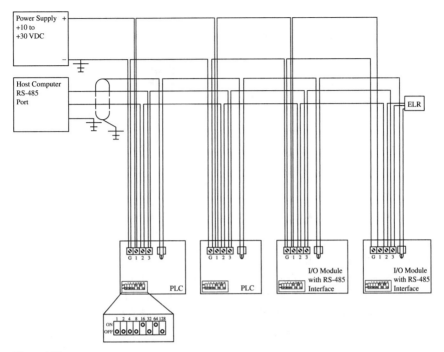

Figure 4-6b

good tape—not the 12 rolls for a dollar stuff. I like to terminate the wire with a crimp-type wire nut to ensure that the connection will not vibrate loose.

Module address Each module in the RS-485 multipoint network must have a unique address. This is usually done with a dip or rotary switch. Figure 4-6b is a dip switch, which sets the module address. Care should be taken to set the module to the proper address or the module will not be accessible from the network. Setting two modules to the same address will disable both modules. The method used to set the address is called Binary Coded Decimal (BCD). Each switch is assigned a value of 1, 2, 4, 8, 16, 32, 64, or 128. You can set a combination of these numbers to any address from 1 to 256; for example, in Fig. 4-6b the switches are set to ON and the values the modules are set to are added—16 + 64 = 80, 2 + 16 = 18, 8 + 32 = 40, and 2 + 4 = 6. Typically the addresses are not random. A wiring diagram, print, or module printout will have a specific address to use to set the correct address for each module. Care should be taken to read the switch. The values can run from right to left or left to right. The ON setting can be on the top or the bottom of the switch.

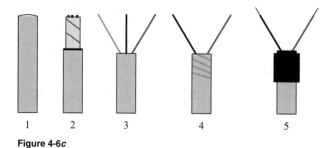

Figure 4-6c

Figure 4-6c is a method for terminating a shielded twisted-pair cable.

1. Shielded twisted-pair cable.
2. Strip the cable, keeping the stripped wire as short as possible, being careful not to nick the wires and expose the shield. Keep the stripped wire away from power conductors to minimize noise.
3. Remove the shield and separate the wires. In Fig. 4-6c the light gray wire is the shield, the black wire is the R (receive data), and the dark gray wire is the T (transmit data).
4. Twist the shield back around the wire to keep it from grounding or shorting.
5. Insulate the shield. I like to use heat shrink tubing. If that is not possible, then wrap tape around the shield, insulating the shield. If you use tape, then use the good tape—it is important that this does not come loose with heat or humidity.

Figure 4-6d is a method for terminating a shielded twisted-pair cable for an RS-485 multipoint network.

1. Shielded twisted-pair cables.
2. Strip each of the cables, keeping the stripped wire as short as possible, being careful not to nick the wires and expose the shield. Keep the stripped wire away from power conductors to minimize noise.
3. Remove the shield and separate the wires. In Fig. 4-6d the light gray wire is the shield, the black wire is the R (receive data), and the dark gray wire is the T (transmit data).
4. Join the cables and twist the shield together.
5. Twist the shields back around the wire to keep it from grounding or shorting.

Monitoring Systems   119

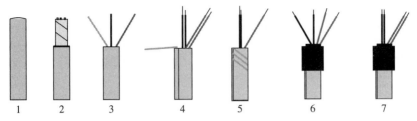

Figure 4-6d

6. Insulate the shield. I like to use heat shrink tubing. If that is not possible, then wrap tape around the shield, insulating the shield. If you use tape, then use good tape—it is important that this does not come loose with heat or humidity.
7. Twist the stripped end of the wires and connect them to the connector.

Figure 4-7 is an Allen Bradley 837E temperature sensor with 4- to 20-mA output; an Allen Bradley MicroLogix 1100 PLC with RS-232, RS-485, and Ethernet; an Allen Bradley 700SE solid-state relay; and a DGH Corporation D1112 RS-485 I/O module with analog input and

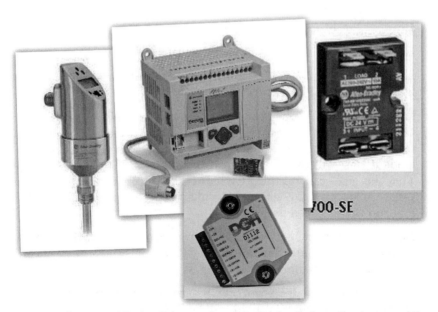

Figure 4-7 *Courtesy of Rockwell Automation, Inc.; for detailed specifications, see Allen Bradley at www.ab.com. Courtesy of DGH Corporation; see detailed specifications at www.dghcorp.com.*

digital output. Using the DGH Corporation modules, you can have up to 124 modules on one RS-485 communication line. This will require repeaters every 32 modules. This may change from manufacturer to manufacturer; check the specification of the modules to determine the maximum number and configuration per line.

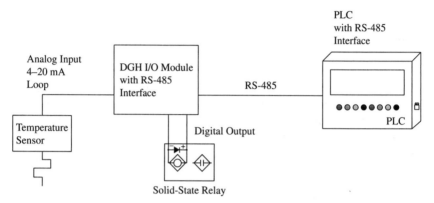

Figure 4-7a

Shown in Fig. 4-7b:

Line 1 Power for SSR1.

Line 2 Power for DGH1.

Line 3 Analog input of 4 to 20 mA to DGH1.

Line 4 Power to temperature sensor. Output from sensor of 4 to 20 mA to input of DGH1.

Line 5 Normally Open contact SSR1, CR1 coil. When the temperature sensor is in the preset range, the DGH1 module outputs a signal to the digital output, which energizes the SSR1, closes the contact, and holds it closed until the temperature is outside of the preset range. This energizes control relay CR1. DGH1 will communicate the temperature to the PLC over the RS-485 network and the PLC can react and run the required program.

Line 6 Normally Open contact CR1, coil motor starter M1, Normally Closed overload contact of M1. When CR1 is energized, the motor will start the blower motor.

Line 7 Normally Open contact CR1, green pilot light. When CR1 is energized, the green pilot light will be illuminated.

Line 8 Normally Closed contact CR1. The red pilot light will be illuminated until CR1 is energized.

Monitoring Systems 121

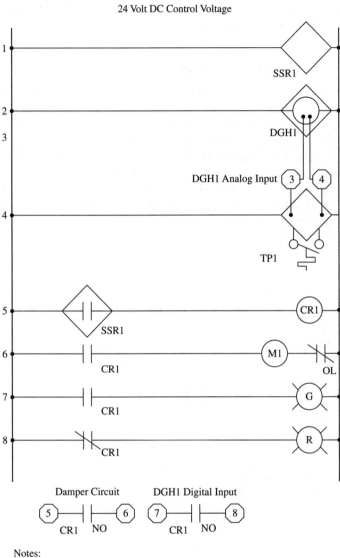

Figure 4-7b

Output contacts

Normally Open contact terminals 5 and 6. When CR1 is energized, the contact will close and open the dampers.

Normally Open contact terminals 7 and 8. When CR1 is energized, the contact will close and close the digital input to DGH1, which will communicate to the PLC over the RS-485 network, and the PLC can react and run the required program.

Modbus Introduction and Specifications

Modbus was introduced in 1979 when PLC manufacturer Modicon, now a brand of Schneider Electric, published the Modbus communication interface for a multipoint network based on a master/client architecture. Simply put, it is a language that devices (nodes) use to communicate with the host system. The original Modbus protocol ran on RS-232, but most later Modbus implementations used RS-485 and Ethernet TCP/IP because it allowed longer distances, higher speeds, and a true multipoint network. In a short time vendors implemented the Modbus messaging system in their devices, and Modbus became the standard for industrial communication networks.

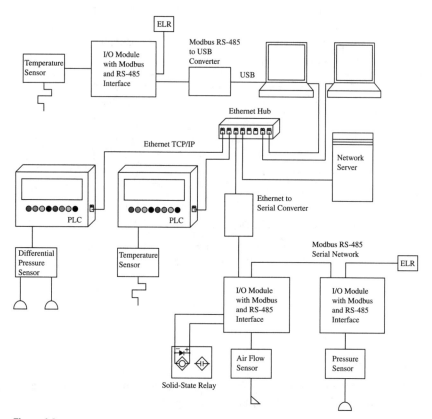

Figure 4-8

Monitoring Systems   123

The Modbus standard has flexibility and is easy to implement. Not only are intelligent devices such as microcontrollers, PLCs, etc., able to communicate with Modbus, but also many intelligent sensors and equipment are equipped with a Modbus-compatible interface to send their data to host systems. With the use of a PLC like Allen Bradley Micrologix 1100 or converters like the units offered by DGH Corporation, it is possible to communicate the Modbus Protocol from one topology to another. This allows a large number of nodes to communicate on RS-485 and connect to the host computers using an Ethernet TCP/IP network. Modbus is a language that is used for the communications of data from node to node and node to host system across many network topologies. It is not a hardware module, topology, or wiring method.

Figure 4-8a is an Allen Bradley 837E temperature sensor with 4- to 20-mA output; an Allen Bradley MicroLogix 1100 PLC with RS-232, RS-485, and Ethernet TCP/IP; an Allen Bradley 700SE solid-state relay; and a DGH Corporation D1112M RS-485 I/O module with analog input and digital output. Using the DGH Corporation modules, you can have up to 247 modules on one RS-485 communication line. This will require repeaters every 32 modules. This may change from manufacturer to

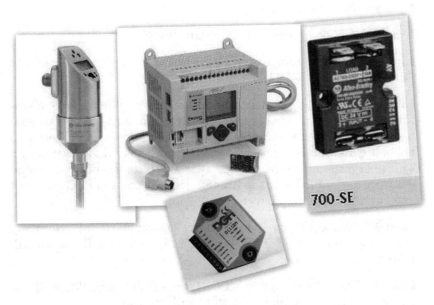

Figure 4-8a *Courtesy of Rockwell Automation, Inc.; for detailed specifications, see Allen Bradley at www.ab.com. Courtesy of DGH Corporation; see detailed specifications at www.dghcorp.com.*

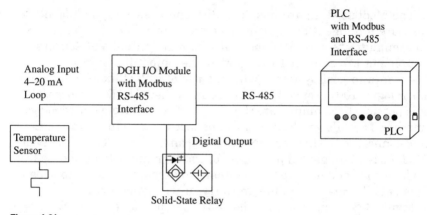

Figure 4-8b

manufacturer; check the specification of the modules to determine the maximum number and configuration per line.

Shown in Fig. 4-8c:

Line 1   Power for SSR1.

Line 2   Power for DGH1.

Line 3   Analog input of 4 to 20 mA to DGH1.

Line 4   Power to temperature sensor. Output from sensor of 4 to 20 mA to input of DGH1.

Line 5   Normally Open contact SSR1, CR1 coil. When the temperature sensor is in the preset range, the DGH1 module outputs a signal to the digital output which energizes the SSR1, closes the contact, and holds it closed until the temperature is outside of the preset range. This energizes control relay CR1. DGH1 will communicate the temperature to the PLC over the RS-485 network and the PLC can react and run the required program.

Line 6   Normally Open contact CR1, coil motor starter M1, Normally Closed overload contact of M1. When CR1 is energized, the motor will start the blower motor.

Line 7   Normally Open contact CR1, green pilot light. When CR1 is energized, the green pilot light will be illuminated.

Line 8   Normally Closed contact CR1, red pilot light will be illuminated until CR1 is energized.

Monitoring Systems 125

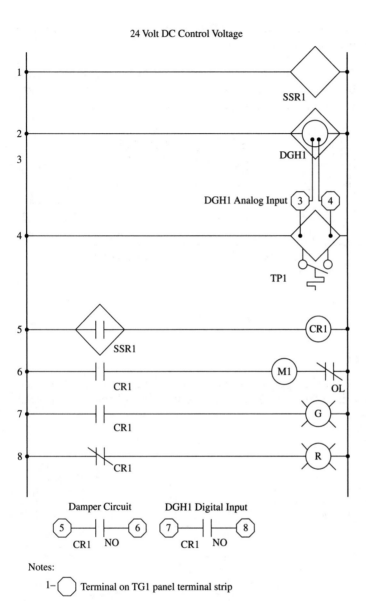

Figure 4-8c

Output contacts

Normally Open contact terminals 5 and 6. When CR1 is energized, the contact will close and open the dampers.

Normally Open contact terminals 7 and 8. When CR1 is energized, the contact will close and close the digital input to DGH1, which will communicate to the PLC over the RS-485 network, and the PLC can react and run the required program.

Figure 4-8d is an example of a typical RS-485 multipoint network with the Modbus protocol. The CRAC units can communicate their settings to the host computer using the Modbus protocol.

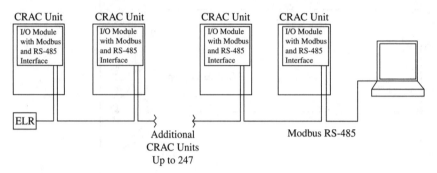

Figure 4-8d

Kilowatt Meter

Figure 4-8e shows the typical use of a monitoring network. It is used here to collect data from the kilowatt meter and send them to the host computer.

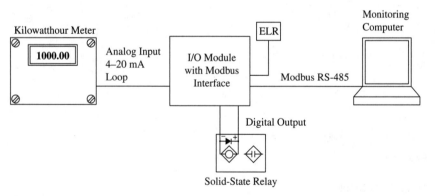

Figure 4-8e

Figure 4-8f shows the use of current and control transformers to collect the data needed by the kilowatt meter.

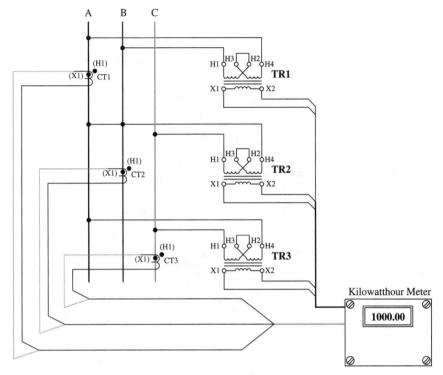

Figure 4-8f

Ethernet Introduction and Specifications

Ethernet is a computer networking technology for local-area networks (LANs). Ethernet is the most widely used network topology. Typical network speeds are 10 to 100 Mbit/s. The most commonly used configuration is 10 BASET 10 Mbps which runs over four wires (two twisted pairs) on CAT3 or CAT5 cable. As shown in Fig. 4-9, a hub or switch sits in the middle of the network and has a port for each node. This configuration is also used for 100 BASET 100 Mbit/s and gigabit Ethernet a gigabit per second. The Ethernet network uses TCP or TCP/IP. TCP is Transmission Control Protocol and IP is Internet Protocol.

Ethernet is typically used for the backbone of the network. The high speeds of Ethernet are not required for monitoring applications. The cost of the Ethernet port prenode network makes it cost-prohibitive for the data collection part of the network. It is more cost-effective to use RS-232 or RS-485 than to convert to Ethernet to join the network.

A commonly used method to convert from RS-232 or RS-485 is to use a PLC such as the Allen Bradley Micrologix 1100 or modules like the

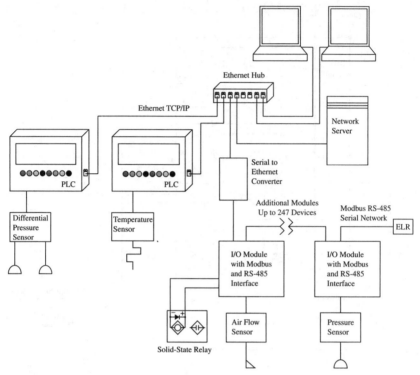

Figure 4-9

A3000 from DGH Corporation. These methods are pass-through devices so that the protocol used on the serial network will be passed to the host network. This allows them to use ASCII or Modbus protocols. (See Fig. 4-9a.)

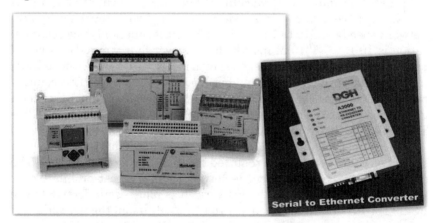

Figure 4-9a *Courtesy of Rockwell Automation, Inc.; for detailed specifications, see Allen Bradley at www.ab.com. Courtesy of DGH Corporation; see detailed specifications at www.dghcorp.com.*

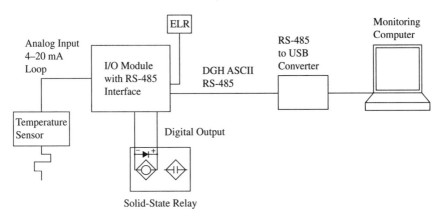

Figure 4-10

Universal Serial Bus

Universal serial bus (USB) is a serial bus standard for the connecting of devices to a host computer. USB was designed to replace serial, parallel, and small computer system interface (SCSI). It allows many peripherals to be connected using a standardized interface socket and standard drivers. USB also allows for the ability to hot-swap devices, connecting and disconnecting devices without rebooting the computer or turning off the device. The USB 2.0 specification has a data transfer rate up to 480 Mbit/s.

There are means to convert from RS-485 to USB such as the one in Fig. 4-10a from DGH Corporation.

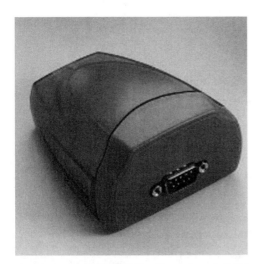

Figure 4-10a USB-COMi. *Courtesy of DGH Corporation; see detailed specifications at www.dghcorp.com.*

Chapter 5

Terminology and Definitions

Alternating Current The term *alternating current* refers to a current that reverses at regular recurring intervals of time and that has alternately positive and negative values of equal amplitude.

Ammeter Device that measures the current flow in amperes in a circuit. An ammeter is connected in series with the load.

Ampere (A) The unit of measurement of electric current flow. If a 1-Ω resistance is connected to a 1-V source, 1 A will flow. Ohm's law states that $I = E/R$, where I is amperage, E is voltage, and R is resistance.

Anode The positive pole of a DC device, or preferably the path by which the current passes out and enters the device on its way to the other pole, opposite of the cathode.

Branch Circuit The circuit conductors between the final overcurrent device protecting the circuit and the device(s).

Calorie The calorie is a pre–International System of Units metric unit of energy. The unit was first defined by Professor Nicolas Clement in 1824 as a unit of heat.

Capacitance Measure, in farads, of the opposition to voltage changes in an AC circuit, causing voltage to lag behind current. Capacitance is the ability of a body to hold an electrical charge. Capacitance is also a measure of the amount of electrical charge stored (or separated) for a given electrical potential.

Capacitive Reactance The opposition to the flow of current in a alternating or pulsating current.

Capacitor A capacitor or condenser is a passive electronic or electrical component consisting of a pair of conductors separated by a dielectric. When a voltage potential difference occurs between the conductors, an electric field occurs in the dielectric. This field can be used to store energy, to resonate with a signal, or to link electrical and mechanical forces.

Running capacitors are used in starting winding to increase the running torque of the motor. Starting capacitors are used in starting winding to increase the starting torque of the motor.

Circuit A complete path through which an electric current can flow.

Circuit Breaker A device designed to open and close a circuit by nonautomatic or automatic means and to open the circuit automatically on a predetermined overcurrent without injury to itself, when properly applied within its rating. Circuit breakers can be reset. They can be tripped remotely with a shunt trip.

Circuit (Series) A circuit supplying energy to a number of devices connected in series. The same current passes through each device, and the voltage is divided between the devices in completing its path to the source of supply.

Closed Circuit A closed path formed by the interconnection of electrical or electronic components through which an electric current can flow.

Coil An assemblage of successive windings of a conductor. A winding consisting of one or more insulated conductors surrounded with insulation and arranged to interact with or produce a magnetic field.

Conductance The measurement of the ease by which electricity flows through a substance. Conductance is measured in mhos because it is the opposite of resistance which is measured in ohms.

Conductor An electrical path that offers comparatively low resistance. A wire or combination of wires not insulated from one another, suitable for carrying a single electric current. Conductors may be classed with respect to their conducting power as (*a*) good (gold, silver, copper, aluminum, zinc, brass, platinum, iron, nickel, tin, lead); (*b*) fair (charcoal and coke, carbon, acid solutions, seawater, saline solutions, metallic ores, living vegetable substances, moist earth); (*c*) partial (water, the body, flame, linen, cotton, mahogany, pine, rosewood, lignum vitae, teak, marble).

Coulomb A unit of electrical charge; the quantity of electricity passing in 1 s with a steady current of 1 A.

Counter EMF The total opposition to the flow of current in an alternating current (AC) circuit.

Cross-Circuit Any accidental contact between electric conductors or wires.

Current (I) The movement of electrons through a conductor; measured in units of amperes.

Cycle A complete reversal of current in an alternating circuit, passing through a complete set of changes or motions in opposite directions, from zero to a rise to maximum amplitude, to a return to zero, to a rise to maximum amplitude in the other direction, and to another return to zero. One complete positive and one complete negative alternation of current or voltage.

Deci A Latin prefix often used with a physical unit to designate a quantity one-tenth ($1/10$) of that unit. 1 deciampere = 0.1 ampere.

Decibel (dB) The *decibel (dB)* is a logarithmic unit of measurement that expresses the magnitude of a physical quantity (usually power or intensity) relative to a specified or implied *reference level*. Usually it is used for sound or signal strength. Since it expresses a ratio of two quantities with the same unit, it is a dimensionless unit. A decibel is one-tenth ($1/10$) of a bel (B).

Deflection The angle distance by which one line or unit departs from another.

Deka A Greek prefix often used with a physical unit denoting a factor of 10^1 or 10. 1 dekaampere = 10 amperes.

Diagram A geometric representation or drawing that illustrates the principles of application of a mechanism or circuit.

Diode A two-electrode electronic device containing an anode and a cathode. Diodes are used as rectifiers, detectors, and light emitters (LEDs).

Direct Current In direct current, the electric charges flow in a constant direction, positive with respect to ground, distinguishing it from alternating current (AC).

Dissipation The loss of electric energy as in the form of heat.

Drop The voltage drop developed across a resistance due to current flowing through it.

E The standard symbol for voltage; short for electromotive force.

Earth The ground reference considered as a grounding medium for an electrical circuit.

Electrical Horsepower By definition, exactly 746 W at 100 percent efficiency.

Electrical Units In an electrical system, electrical units consist of the volt, ampere, ohm, watt, watthour, coulomb, henry, mho, joule, and farad.

Electric Circuit The path (whether metallic or nonmetallic) of an electric current typically consisting of wire(s) and device(s).

Electrician A tradesman specializing in electrical wiring of buildings and related equipment. A tradesman typically begins as an apprentice, working for and learning from a journeyman or master, and after a number of years is released from his or her master's service as a journeyman.

Electricity *Electricity* [from the Greek word ηλεκτρον (electron), meaning amber, and finally from the New Latin *ēlectricus,* "amber-like"] is a general term that encompasses a variety of phenomena resulting from the presence and flow of electric charge. These include many easily recognizable phenomena such as lightning and static electricity, but in addition, less familiar concepts such as the electromagnetic field and electromagnetic induction.

Electrocution The annihilation of life by means of electric current.

Electromagnet A type of magnet in which the magnetic field is produced by the flow of electric current in a conductor. The magnetic field disappears when the flow current ceases.

Electromotive Force (EMF)   The flow of current in an alternating current (AC) circuit. An energy-charge relation that results in electric pressure (voltage), which produces or tends to produce charge flow.

Electron   A subatomic particle that carries a negative electrical charge.

Energy Efficiency   The efficiency of an entity (a device, component, or system) in electronics and electrical circuits. Defined as useful power output divided by the total electrical power consumed. Efficiency = useful power output/total power input.

Farad (F)   The unit of electrostatic capacity in the electromagnetic system. A condenser or capacitor is said to have a capacity of 1 farad (1 F) if it will absorb 1 coulomb (1 C) (1 A/s) of electricity when subjected to a pressure of 1 V. The unit of capacitance.

Faraday Effect   A discovery made by Michael Faraday. In physics, a magneto-optical phenomenon, or an interaction between light and magnetic field in a medium.

Fathom   A measure of length equal to 2 yards (6 ft), used chiefly in taking soundings, measuring cordage, etc. Based on the distance between the fingertips of a man's outstretched arms.

Fiber Optics   A glass or plastic fiber that carries light along its length.

Fluorescence   A luminescence that is mostly found as an optical phenomenon in cold bodies, in which the molecular absorption of a photon triggers the emission of a photon with a longer (less energetic) wavelength. That property by which certain solids and fluids become luminous under the influence of radiant energy.

Force   An elementary physical cause capable of modifying the motion of a mass; it is measured in newtons.

Formula   In mathematics and in the sciences, a formula (plural: formulae, formulæ, formulas) is a concise way of expressing information symbolically. A rule or principle expressed in algebraic language.

Frequency   The number of periods occurring in the unit of time periodic process, such as in the cycles of electric charge. The number of complete cycles per second existing in any form of wave motion, such as the number of cycles per second of an alternating current.

Fuse A strip of wire or metal inserted in series with a circuit which, when it carries an excess of current over its rated capacity, will burn out, thus protecting the circuit's other components from damage due to excessive current.

Galvanometer A type of ammeter—an instrument for detecting and measuring electric current. It is an analog electromechanical transducer that produces a rotary deflection, through a limited arc, in response to electric current flowing through its coil. The term has been expanded to include uses of the same mechanism in recording, positioning, and servomechanism equipment.

Generator A device that converts mechanical energy to electrical energy, generally using electromagnetic induction.

Ground The term *ground* or *earth* has several meanings depending on the specific application areas. Ground may be the reference point in an electrical circuit from which other voltages are measured, a common return path for electric current (*earth return* or *ground return*), or a direct physical connection to the earth.

Grounded The term *grounded* refers to the act of connecting the electrical circuit to a ground or earth reference. This may be done for safety purposes during installation or maintenance.

Heat (Electric) The power loss produced in a conductor having electric current flow through it. Electric heaters and stoves use electromagnetic dissipation to produce heat.

Horsepower (hp) Unit used to express rate of work, or power. 1 horsepower (hp) = 746 watts (W) at 100 percent efficiency. Work is done at the rate of 33,000 foot-pounds per minute or 550 foot-pounds per second. The term *horsepower* is being slowly replaced by kilowatt (kW) and megawatt (MW).

I Standard symbol for electric current.

Impedance (Z) The total opposition to the flow of current in an alternating current (AC) circuit at a given frequency; combination of resistance and reactance, measured in ohms.

Inductance (L) The property of an alternating current electrical circuit where the change in the current through that circuit induces a counter electromotive force that opposes the change in current.

Induction The production of current flow in a conductor in a magnetic field.

Input The entrance point in an electrical circuit or device, as in the input to a sensor or motor.

Insulator Also called a *dielectric*; a material that resists the flow of electric current. A device for fastening and supporting a conductor.

Ion An atom or a molecule that has lost or gained one or more electrons, giving it a positive or negative electrical charge.

Joint The connection of electrical conductors so that the union will be good, both mechanically and electrically.

Joule's Law Also known as the *Joule effect*. A physical law expressing the relationship between the heat generated by the current flowing through a conductor. It is named for James Prescott Joule, who studied the phenomenon in the 1840s.

Kilovolt (kV) A unit of measurement equal to 1000 volts. 1000 V = 1 kV.

Kilowatt (kW) Equal to 1000 watts. Typically used to state the power output of engines and the power consumption of tools and machines. A kilowatt is approximately equivalent to 1.34 horsepower (hp). An electric heater with one heating element might use 1 kW. 1000 W = 1 kW.

A distinction should be made between kilowatts, which is the measurement of resistance, and kilovolt amperes reactive, which is the measurement of impedance in an AC circuit.

Leakage The loss of electric current through defects in insulation or other causes.

Loss Power used or expended without accomplishing useful work.

Megavolt A unit of measurement equal to 1 million volts. 1 MV = 1,000,000 V.

Meter An electric measuring instrument such as a voltmeter, ammeter, kilowatthour meter, etc.

Negative The opposite of positive. A potential less than that of another potential. In electrical apparatus, the pole or direction toward which the current flows. A negative value can have a positive sign. It is negative with respect to another potential.

Network In the context of an electrical circuit, a collection of interconnected components. The components are connected in some special manner.

Neutron A subatomic particle with no net electric charge and a mass slightly larger than that of a proton.

Ohm (Ω) Defined as the resistance between two points of a conductor when a constant potential difference of 1 V, applied to these points, produces in the conductor a current of 1 A, the conductor not being the seat of any electromotive force.

$$R = E/I$$

where R = resistance, Ω

E = volts

I = amperes

Open Circuit A circuit in which the electrical continuity has been interrupted.

Output The result of an action or sensor that produces current or voltage in a circuit or device.

P Standard abbreviation for power.

Peak The maximum measurable value of a varying voltage or current.

Peak Current The maximum measurable value of an alternating current.

Period A measurement of time. The time required for a complete cycle of alternating current or voltage; for 60 cycles per second, a period is $1/60$ s.

Photoelectric A phenomenon in which electrons are emitted from matter after the absorption of energy from electromagnetic radiation such as visible light.

Photoelectric Sensor Electronic device recognizing changes in light intensity and converting these changes to a change in output state.

Photon An elementary particle, the quantum of the electromagnetic field, and the basic unit of light and all other forms of electromagnetic radiation. It is also the force carrier for the electromagnetic force.

Positive   This is a relative term. It is positive relative to another potential. A positive value can have a negative sign in respect to a less positive value.

Power (*P*)   Defined as the rate at which electrical energy is transferred by an electric circuit, measured in watts or horsepower. $P = I \cdot E$, where P = power in watts, I = current measured in amperes, and E = volts.

Proton   A subatomic particle with an electric charge of +1 elementary charge. The smallest quantity of electricity that can exist in the free state. A positive charged particle in the nucleus of an atom.

Quick-Break   A switch, circuit breaker, or contact that has a high contact opening speed.

R   Standard symbol for resistance.

Reactance (*X*)   Opposition to the flow of current in an AC circuit by the inductance or capacity of a part; measured in ohms.

Relay   An electromagnetic or solid-state device with contact(s) either Normally Open or Normally Closed, which permits control of current in a circuit.

Resistance (*R*)   The opposition to the flow of current in a circuit that produces heat. Pure resistance can only be found in a DC circuit.

Resistor   A device that introduces resistance to a circuit, producing a voltage drop in its terminals in proportion to the current.

Semiconductor   A solid material that has an electrical conductivity between those of a conductor and an insulator; it can vary over that wide range either permanently or dynamically.

Series Circuit   A circuit supplying energy to a number of loads connected in series. The same current passes through each load, and the voltage drops in proportion to the resistance of the device in completing its path to the source of supply.

Series Parallel Circuit   An electric circuit containing parallel connected devices and serial connected devices.

Short Circuit   A fault in an electric circuit or apparatus due to human error or an imperfection in the insulation. The current flows through a path not intended.

Shunt A device that short-circuits an electrical circuit.

Shunt Trip A coil that allows the remote disconnection of electrical power by mechanically tripping the circuit breaker.

Solenoid A coil or transducer that, when an electric current passes through it, causes a linear motion in a mechanical operator.

Steady Current An electric current of constant value measured in amperes.

Transformer A device that transfers electrical energy from one circuit to another through inductively coupled electrical conductors. A changing current in the first circuit (the *primary*) creates a changing magnetic field. This changing magnetic field induces a changing voltage in the second circuit (the *secondary*). This effect is called *mutual induction*.

Transformer (Current) A transformer that is used to measure the current in a circuit. Care should be taken with the secondary of the current transformer. The secondary should not be disconnected from the load while current is flowing. The transformer will attempt to continue driving current across the effectively infinite impedance. This will produce a high voltage across the secondary, which can be in the range of several kilovolts and can cause arcing. This voltage will compromise your safety and permanently affect the accuracy of the transformer.

Transistor A semiconductor device commonly used to amplify or switch electronic signals.

Unit of Current The standard unit of measurement for current is the ampere, which is the current produced by 1 volt in a circuit having a resistance of 1 ohm.

Unit of Pressure Pressure or volts in an electrical circuit that will produce a current of 1 ampere against a resistance of 1 ohm.

Unit of Resistance The unit of measurement is the ohm, which is the resistance that permits a flow of 1 ampere when the pressure is 1 volt.

Volt (V) The unit of measurement of electric pressure; the pressure that will produce a current of 1 ampere against a resistance of 1 ohm.

Voltage Drop The drop of voltage in an electric circuit due to the resistance of the conductor.

Watt (W) The unit of measurement of electrical power, being the amount of energy expended per second by an unvarying current of 1 ampere and 1 volt.

X Standard symbol for reactance.

***Y* Connection** This method of transformer connection consists of connecting primaries and/or secondaries in a star grouping.

Z Standard symbol for impedance.

Appendix A

Motor Control—3 Phase

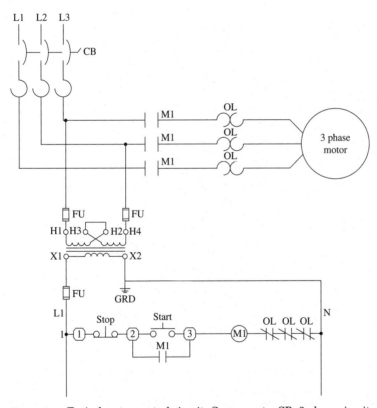

Figure A-1 Typical motor control circuit. Components: CB, 3-phase circuit breaker with thermal trip units; OL, motor thermal overloads; FU, fuse rated to protect the control circuit; control transformer configured for high-voltage input low-voltage output; Stop, Normally Closed push button; Start, Normally Open push button; M1, low-voltage control coil part of the motor controller.

Appendix B
Ladder Diagrams

Appendix B

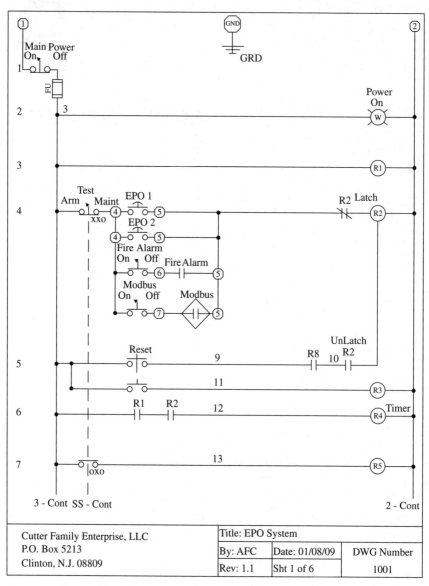

Figure B-1 Emergency power off system (sheet 1 of 6).

Ladder Diagrams 147

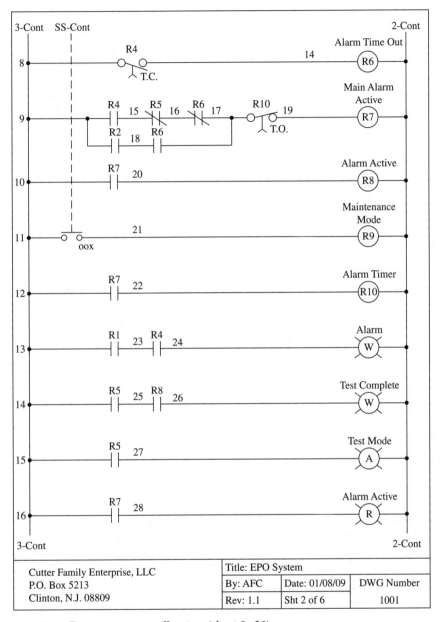

Figure B-2 Emergency power off system (sheet 2 of 6).

148 Appendix B

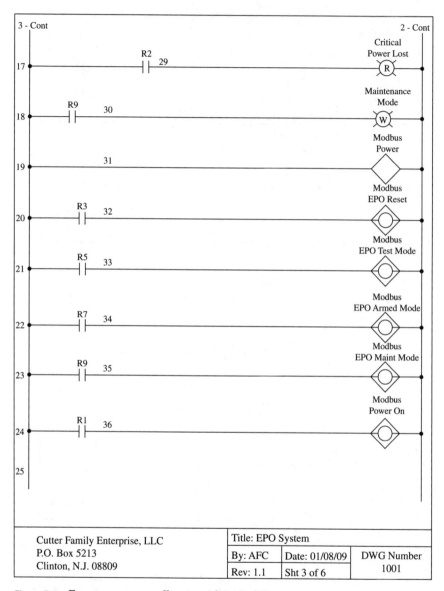

Figure B-3 Emergency power off system (sheet 3 of 6).

Ladder Diagrams 149

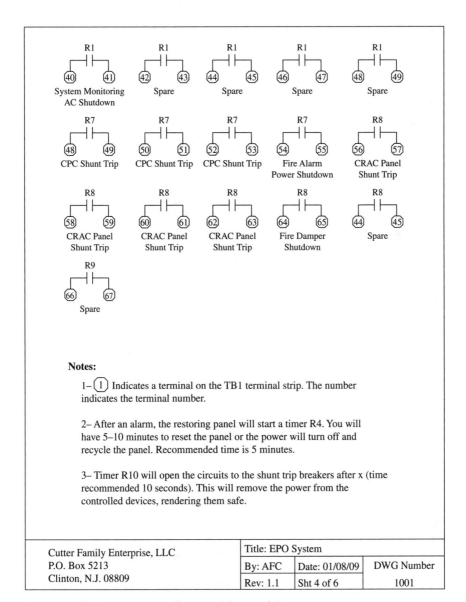

Figure B-4 Emergency power off system (sheet 4 of 6).

Appendix B

Bill of Materials			
Item	**Qty**	**Manufacturer**	**Part Number**
4-Pole Relay	8	Allen-Bradley	700-N400A
4-Pole Time Delay Relay	2	Allen-Bradley	700-NT400A
4-Pole Contact Front Deck	10	Allen-Bradley	700-NA00
4-Position Relay Rack	2	Allen-Bradley	700-NP4
Terminal Block	55	Allen-Bradley	1492-CA2
Terminal Rail	1	Allen-Bradley	1492-CA2175(88/Rail)
Fuse Block	1	Allen-Bradley	1492-H6
2-Position Keyed Selector Switch	2	Allen-Bradley	800T-H33D1
3-Position Keyed Selector Switch	1	Allen-Bradley	800T-J44A
EPO Mushroom EPO Push Button	2	Allen-Bradley	800T-E15M6A
2-Pole Push Button Black Reset	1	Allen-Bradley	800T-A2A
Pilot Light White	3	Allen-Bradley	800T-QH10W
Pilot Light Red	2	Allen-Bradley	800T-QH10R
Pilot Light Amber	1	Allen-Bradley	800T-QH10A
Wireway		Panduit	G1LG6
Wireway Cover		Panduit	C11G6

Cutter Family Enterprise, LLC
P.O. Box 5213
Clinton, N.J. 08809

Title: EPO System
By: AFC | Date: 01/08/09 | DWG Number
Rev: 1.1 | Sht 5 of 6 | 1001

Figure B-5 Emergency power off system (sheet 5 of 6).

Ladder Diagrams   151

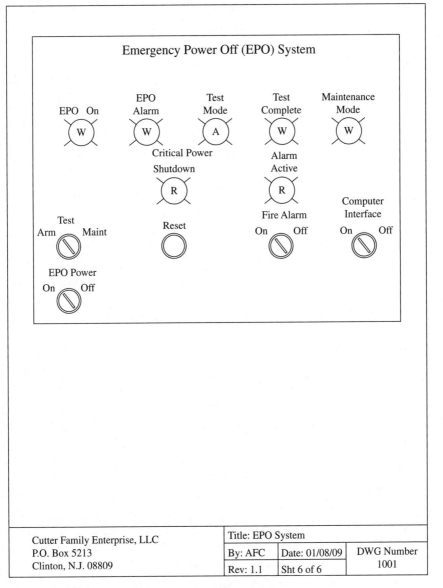

Figure B-6 Emergency power off system (sheet 6 of 6).

152 Appendix B

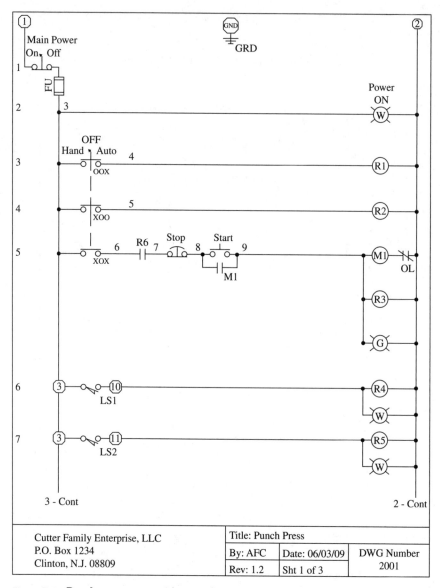

Figure B-7 Punch press system (sheet 1 of 3).

Ladder Diagrams 153

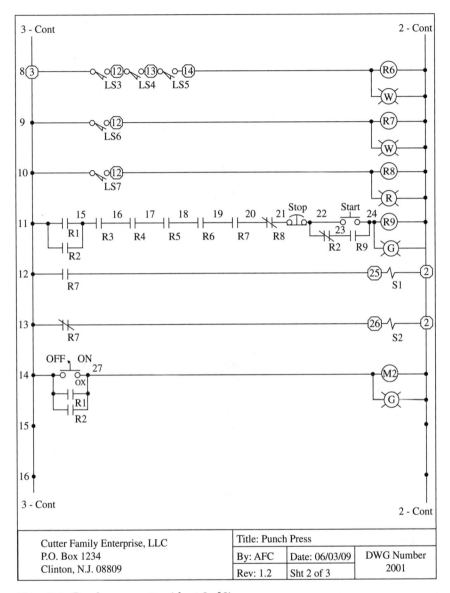

Figure B-8 Punch press system (sheet 2 of 3).

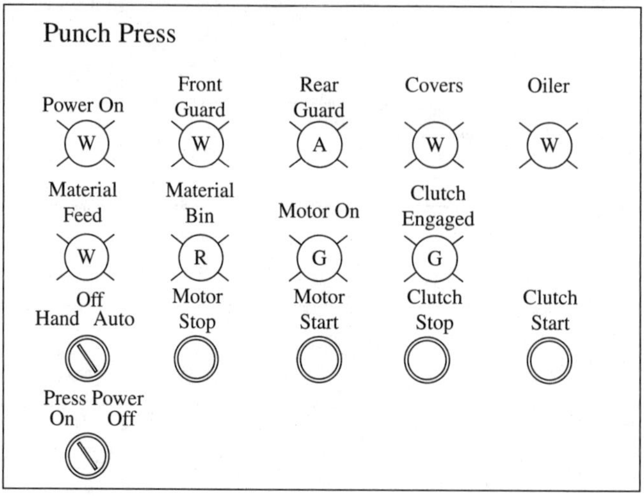

Punch Press

```
                Front        Rear
    Power On    Guard        Guard       Covers      Oiler
      (W)        (W)          (A)         (W)         (W)

    Material   Material                  Clutch
     Feed       Bin       Motor On      Engaged
      (W)        (R)         (G)          (G)

       Off      Motor       Motor       Clutch      Clutch
    Hand Auto    Stop       Start        Stop       Start
       ⊘         ○           ○           ○           ○

    Press Power
     On   Off
       ⊘
```

Notes:

1 - LS1 - Limit Switch mounted on Front Guard
2 - LS2 - Limit Switch mounted on Rear Guard
3 - LS3 - Limit Switch mounted on Motor Cover Guard
4 - LS4 - Limit Switch mounted on Belt Cover Guard
5 - LS5 - Limit Switch mounted on Clutch Cover Guard
6 - LS6 - Limit Switch mounted on Material Feed
7 - LS7 - Limit Switch mounted on Material Output Bin
8 - R1 - Relay Coil Engaged in Auto Mode
9 - R2 - Relay Coil Engaged in Hand Mode
10 - R3 - Relay Coil Engaged when Motor 1 is Running
11 - R4 - Relay Coil Engaged when Front Guard is Closed
12 - R5 - Relay Coil Engaged when Rear Guard is Closed
13 - R6 - Relay Coil Engaged when Belt Cover, Motor Cover, and Clutch are Closed
14 - R7 - Relay Coil Engaged when Material is in Feeder
15 - R8 - Relay Coil Engaged when Material Bin is Full
16 - R9 - Relay Coil Engaged when Clutch is Running
17 - S1 - Clutch Solenoid
17 - S2 - Brake Solenoid

Cutter Family Enterprise, LLC P.O. Box 1234 Clinton, N.J. 08809	Title: Punch Press		
	By: AFC	Date: 06/03/09	DWG Number
	Rev: 1.2	Sht 3 of 3	2001

Figure B-9 Punch press system (sheet 3 of 3).

Appendix C

DGH Corporation Modules

**D1000 and D2000 SERIES
SENSOR TO COMPUTER INTERFACE MODULES**

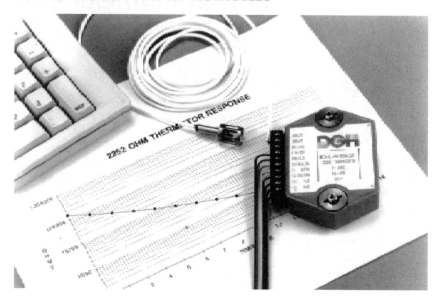

D1000 and D2000 FEATURES
- Complete sensor to RS-485 or RS-232C interface.
- ASCII format command/response protocol.
- 500V rms analog input isolation.
- 15 bit measurement resolution.
- Continuous self-calibration; no adjustments of any kind.
- Programmable digital filter.
- Digital limit setting and alarm capability.
- Digital inputs and outputs connect to solid state relays.
- Events counter to 10 million.
- Requires +10V to +30Vdc unregulated supply.
- Transient suppression on RS-485 communications lines.
- Screw terminal plug connectors supplied.

D2000 PROGRAMMABLE FEATURES
Provides intelligent features not found in the D1000.
- ASCII output scaled to desired engineering units.
- User programmable nonlinear transfer function.
- Straight-line segment approximation: up to 24 segments.

APPLICATIONS
- Process monitoring and control
- Remote data logging to any host computer
- Product testing
- Direct connection to modems

Figure C-1 *Courtesy of DGH Corporation. See detailed specifications at www.dghcorp.com.*

DGH Corporation Modules   157

D1000 and D2000 SPECIFICATIONS (typical at +25°C and nominal power supply unless otherwise noted)

Analog
- Single channel analog input.
- Maximum CMV, input to output at 60Hz: 500V rms.
- Leakage current, input to output at 115Vrms, 60Hz: <2µA rms.
- 15 bit measurement resolution.
- 8 conversions per second.
- Autozero & autocalibration—no adjustment pots.

Digital
- 8-bit CMOS microcomputer.
- Digital scaling, linearization and calibration.
- Nonvolatile memory eliminates pots and switches.

Digital filtering
- Small and large signal with user selectable time constants from 0 to 16 seconds.

Events counter
- Up to 10 million positive transitions at 60Hz max., filtered for switch debounce.

Digital inputs
- Voltage levels: ±30V without damage.
- Switching levels: High, 3.5V min., Low, 1.0V max.
- Internal pull up resistors for direct switch input.

Digital outputs
- Open collector to 30V, 30mA max. load.

Alarm outputs
- HI/LO limit checking by comparing input values to downloaded HI/LO limit values stored in memory.
- Alarms: latching (stays on if input returns to within limits) or momentary (turns off if input returns to within limits).

Communications
- Communications in ASCII via RS-232C, RS-485 ports.
- Selectable baud rates: 300, 600, 1200, 2400, 4800, 9600, 19200, 38400.
- NRZ asynchronous data format; 1 start bit, 7 data bits, 1 parity bit and 1 stop bit.
- Parity: odd, even, none.
- User selectable channel address.
- ASCII format command/response protocol.
- Up to 124 multidrop modules per host serial port.
- Communications distance up to 4,000 feet (RS-485).
- Transient suppression on RS-485 communications lines.
- Communications error checking via checksum.
- Can be used with "dumb terminal".
- Scan up to 250 channels per second.
- All communications setups stored in EEPROM.

Power
Requirements: Unregulated +10V to +30Vdc, 0.75W max (D1500/D2500, 2.0W max.).
Internal switching regulator.
Protected against power supply reversals.

Environmental
Temperature Range: Operating -25°C to +70°C.
Storage -25°C to +85°C.
Relative Humidity: 0 to 95% noncondensing.

Warranty
12 months on workmanship and material.

D1100/D2100 Voltage Inputs
- Voltages: ±10mV, ±100mV, ±1V, ±5V, ±10V, ±100Vdc.
- Resolution: 0.01% of FS (4 digits).
- Accuracy: ±0.02% of FS max.
- Common mode rejection: 100dB at 50/60Hz.
- Zero drift: ±1 count max (autozero).
- Span tempco: ±50ppm/°C max.
- Input burnout protection to 250Vac.
- Input impedance: $\leq \pm 1V$ input = 100MΩ min.
  $\geq \pm 5V$ input = 1MΩ min.
- 1 Digital input/Event counter, 2 Digital outputs.

D1200/D2200 Current Inputs
- Currents: ±1mA, ±10mA, ±100mA, ±1A, 4-20mAdc.
- Resolution: 0.01% of FS (4 digits), 0.04% of FS (4-20mA).
- Accuracy: ±0.02% of FS, 0.04% of FS (4-20mA).
- Common mode rejection: 100dB at 50/60Hz.
- Zero drift: ±1 count max (autozero).
- Span tempco: ±50ppm/°C max. (±1A = ±80 ppm/°C max.)
- Voltage drop: ±0.1V max.
- 1 Digital input/Event counter, 2 Digital outputs.

D1300 Thermocouple Inputs
- Thermocouple types: J, K, T, E, R, S, B, C (factory set).
- Ranges: J = -200°C to +760°C   B = 0°C to +1820°C
  K = -150°C to +1250°C   S = 0°C to +1750°C
  T = -200°C to +400°C   R = 0°C to +1750°C
  E = -100°C to +1000°C   C = 0°C to +2315°C
- Resolution: ±1°.
- Overall Accuracy (error from all sources) from 0 to +40°C ambient: ±1.0 °C max (J, K, T, E).
  ±2.5 °C max (R, S, B, C)(300°C TO FS).
- Common mode rejection: 100dB at 50/60Hz.
- Input impedance: 100MΩ min.
- Lead resistance effect: <20µV per 350Ω.
- Open thermocouple indication.
- Input burnout protection to 250Vac.
- User selectable °C or °F.
- Overrange indication.
- Automatic cold junction compensation and linearization.
- 2 Digital inputs, Event counter, 3 Digital outputs.

D1400 RTD Inputs
- RTD types: α = .00385, .00392, 100Ω at 0°C,
  .00388, 100Ω at 25°C.
- Ranges: .00385 = -200°C to +850°C.
  .00392 = -200°C to +600°C.
  .00388 = -100°C to +125°C.
- Resolution: 0.1°.
- Accuracy: ±0.3°C.
- Common mode rejection: 100dB at 50/60Hz.
- Input connections: 2, 3, or 4 wire.
- Excitation current: 0.25mA.
- Lead resistance effect: 3 wire - 2.5°C per Ω of imbalance.
  4 wire - negligible.
- Max lead resistance: 50Ω.
- Input protection to 120Vac.
- Automatic linearization and lead compensation.
- User selectable °C or °F.
- 1 Digital output.

Figure C-1 (*Continued*)

158   Appendix C

D1450 Thermistor Inputs
- Thermistor types: 2252Ω at 25°C, TD Series
- Ranges: 2252Ω = -0°C to +100°C.
  TD = -40°C to +150°C.
- Resolution: 2252Ω = 0.01°C or F.
  TD = 0.1°C or F
- Accuracy: 2252Ω = ±0.1°C.
  TD = ±0.2°C
- Common mode rejection: 100dB at 50/60Hz.
- Input protection to 30Vdc .
- User selectable °C or °F.
- 1 Digital input/ Event counter, 2 Digital outputs.

D1500/D2500 Bridge Inputs
- Voltage Ranges: ±30mV, ±100mV, 1-6Vdc.
- Resolution: 10μV (mV spans).
  0.02% of FS (V span).
- Accuracy: ±0.05% of FS max.
- Common mode rejection: 100dB at 50/60Hz.
- Input protection to 30Vdc .
- Offset Control: Full input range.
- Excitation Voltage: 5V, 8V, 10Vdc, 60mA max.
- Zero drift: ±1μV/°C max.
- Span tempco: ±50ppm/°C max.
- 1 Digital output.

D1600/D2600 Timer and Frequency Inputs
- Input impedance: 1MΩ.
- Switching level: selectable 0V, +2.5V.
- Hysteresis: Adjustable 10mV-1.0V.
- Input protection: 250Vac .
- 1 Digital input/Event counter.

Frequency Input
- Range: 1Hz to 20KHz.
- Resolution: 0.005% of reading + 0.01Hz.
- Accuracy: ±0.01% of reading ±0.01Hz.
- Tempco: ±20ppm/°C.

Timer Input
- Range: 100μs to 30 s.
- Resolution: 0.005% of reading +10μs.
- Accuracy: ±0.01% of reading ±10μs.
- Tempco: ±20ppm/°C.

Event Counter Input
- Input Bandwidth: 60Hz, (optional 20KHz max.).
- Up to 10 million positive transitions.

Accumulator Input
- Input Frequency Range: 1Hz to 10KHz.
- Input Timer Range: 100μs to 30s.
- Pulse Count: Up to 10 million positive transitions.
- Resolution: 0.005% of reading +0.01Hz (Frequency).
  0.005% of reading +10μs (Timer) .
- Accuracy: ±0.01% of frequency reading ±0.01Hz.
  ±0.01% of timer reading ±10μs.
- Tempco: ±20ppm/°C.

Specifications are subject to change without notice

D1700 Digital Inputs/Outputs
D1711, D1712: 15 digital input/output bits.
- User can define any bit as an input or an output.
- Input voltage levels: 0-30V without damage.
- Input switching levels: High, 3.5V min., Low, 1.0V max.
- Outputs: Open collector to 30V, 100mA max. load.
- Vsat: 1.0V max @ 100mA.
- Single bit or parallel I/O addressing.

D1701, D1702: 7 digital inputs and 8 digital outputs.
- Input voltage levels: ±30V without damage.
- Input switching levels: High,3.5V min.,Low,1.0V max.
- Outputs: open collector to 30V, 30mA max. load.
- Vsat: 0.2V max @ 30mA.
- Internal pull up resistors for direct switch input.
- Inputs/Outputs are read/set in parallel.

Specifications are subject to change without notice.

Mechanicals and Dimensions
Case: ABS with captive mounting hardware.
Connectors: Screw terminal barrier plug (supplied).
  Replace with Phoenix MSTB 1.5/10 ST 5.08
  or equivalent.

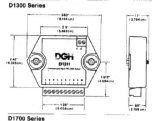
D1300 Series

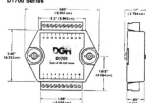
D1700 Series

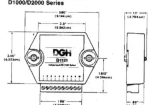
D1000/D2000 Series

NOTE: Spacing for mounting screws = 2.700" (6.858 cm).
Screw threads are 6 X 32.

Figure C-1   *(Continued)*

DGH Corporation Modules 159

GENERAL DESCRIPTION

The D1000 and D2000 Sensor to Computer Modules are a family of complete solutions designed for data acquisition systems based on personal computers and other processor-based equipment with standard serial I/O ports. The modules convert analog input signals to engineering units and transmit in ASCII format to any host with standard RS-485 or RS-232C ports. These modules can measure temperature, pressure, voltage, current and various types of digital signals. The modules provide direct connection to a wide variety of sensors and perform all signal conditioning, scaling, linearization and conversion to engineering units. Each module also provides digital I/O lines for controlling devices through solid state relays or TTL signals. These digital I/O lines along with built-in limit setting capability provide alarm and control outputs.

The modules contain no pots or switches to be set. Features such as address, baud rate, parity, alarms, echo, etc. are selectable using simple commands over the communications port—without requiring access to the module. The selections are stored in nonvolatile EEPROM which maintains data even after power is removed.

The key to the DGH product concept is that the modules are easy to use. You do not need engineering experience in complicated data acquisition hardware. With the DGH modules, anyone familiar with a personal computer can construct a data acquisition system. This modular approach to data acquisition is extremely flexible, easy to use and cost effective. Data is acquired on a per channel basis so you only buy as many channels as you need. The modules can be mixed and matched to fit your application. They can be placed remote from the host and from each other. You can string up to 124 modules on one set of wires by using RS-485 with repeaters.

The D2000 series is an enhanced version of the D1000 series of sensor interfaces. The D2000 series allows the user to scale the output data in any desired engineering units. The D2000 also provides the ability to program nonlinear transfer functions. This feature may be used to linearize nonstandard sensors or to provide outputs in engineering units which are nonlinear functions of the input.

The D2000 can be programmed to approximate square law, root, log, high-order polynomial or any other nonlinear function. The D2000 may also be empirically field-programmed when the exact transfer function is unknown.

The D1000 and D2000 modules are isolated data acquisition systems for real-time distributed processing and control. By distributing computer power to each sensor location, the host computer is unburdened from interpreting data from sensor inputs. Instead of scaling and linearizing sensor data, the host computer can be used more efficiently to scan a greater number of inputs and to provide faster control output.

The D1000 and D2000 are compatible with the DGH D3000 and D4000 series and may be mixed in any combination. The D3000 and D4000 series convert ASCII format input commands to voltage or current output signals.

All modules are supplied with screw terminal plug connectors and captive mounting hardware. The connectors allow system expansion, reconfiguration or repair without disturbing field wiring.

UTILITY SOFTWARE

Complimentary Utility Software is included with each purchase order. The software is compatible with Windows 95, 98, NT 4.0+, 2000 operating systems and distributed on CD-ROM. The Utility Software simplifies configuration of all user-selectable options such as device address, baud rate and filtering constants. The latest version of our software is always downloadable from our web site at www.dghcorp.com.

THEORY OF OPERATION

Each DGH module is a complete single-channel data acquisition system. Each unit contains analog signal conditioning circuits optimized for a specific input type. The amplified sensor signals are converted to digital data with a microprocessor-controlled integrating A/D converter. Offset and gain errors in the analog circuitry are continuously monitored and corrected using microprocessor techniques. The D1000 converts the digital signal data into engineering units using look-up tables. The D2000 converts the digital signal data into engineering units using look-up tables that are customer-programmed. The resultant data is stored in ASCII format in a memory buffer. The modules continuously convert data at the rate of 8 conversions per second and store the latest result in the buffer. The host computer may request data by sending simple ASCII commands to the module. The D1000 will then instantly respond by communicating the ASCII buffer data back to the host. Up to 124 modules may be linked to a single RS-232C or RS-485 host computer port. Each module on a serial line is identified by a unique user-programmable address. This addressing technique allows modules to be interrogated in any order.

DIGITAL INPUTS/OUTPUTS

D1000 and D2000 modules also contain up to three digital outputs and two digital inputs. The digital outputs are open-collector transistor switches that may be controlled by the host computer. These switches may be used to control solid-state relays which in turn may control heaters, pumps and other power equipment. The digital inputs may be read by the host computer and used to sense the state of a remote digital signal. They are ideal for sensing the state of limit or safety switches. Digital I/O capability may be expanded by using the DGH D1700 modules.

Figure C-1 (*Continued*)

EVENT COUNTER

With the exception of D1400 RTD, D1500 and D2500 bridge input modules, every module contains an onboard event counter. The event counter will count up to 10 million transitions that occur on the digital input. The event counter may be read and cleared by the host computer at any time. The counter has many applications where a host computer must read an accumulated count of events. It may be used in production line applications to keep a record of repetitious operations. For applications that only require counting, DGH offers the D1621 and D1622 Event Counter modules. These modules have no analog input but count events up to 10 million at either 60Hz or 20KHz bandwidths.

For applications that require reading and accumulating pulse-type information DGH offers the Accumulator modules. The Accumulators can read both the rate and the total count of a frequency or pulse input signal. They can keep track of power consumption when connected to a power meter or accumulate the output of pulse-type flow meters.

ALARM OUTPUTS

The D1000 and D2000 modules include digital high and low alarm functions. High and low alarm limits may be downloaded into the module by the host computer. The limit data is compared against the analog input data after every A/D conversion. The result of the limit comparison may be read by the host. The high and low limits may also be used to control the digital outputs on the module. The limits may be used to turn on alarms or to shut down a process independent of a host computer. Limit data may be changed at any time with commands from the host computer. Limit values are stored in nonvolatile memory to preserve the values even when module power is removed. Limit data is downloaded in the same engineering units as output data. Alarm outputs may be programmed to be latching to record the occurrence of a single alarm event. Alarm outputs may also be configured to form simple on-off controllers that are independent of the host computer.

USER OPTIONS

To provide maximum flexibility, the D1000 and D2000 offer a variety of user-selectable options including choice of address, baud rate, parity, alarm options, echo, etc. All options are selectable using simple commands over the communications port. All option selections are stored in a nonvolatile EEPROM which maintains data even after power is removed. The modules contain no pots or switches to be set. All options may be changed remotely without requiring access to the module.

DIGITAL FILTER

The D1000 and D2000 options include a unique programmable single pole digital filter. The filter is used to smooth analog data in noisy environments. Separate time constants may be specified for small and large signal changes. Typically a large time constant is specified for small signal changes to filter out noise and provide stable output readings. A smaller time constant may be chosen for large signal changes to provide fast response to such changes.

COMMUNICATIONS

The D1000 and D2000 are designed to be easy to interface to all popular computers and terminals. All communications to and from the module are performed with printable ASCII characters. This allows the information to be processed with string functions common to most high-level languages such as BASIC. For computers that support standard ports such as RS-232C, no special machine language software drivers are necessary for operation. The modules can also be connected to auto-answer modems for long-distance operation without the need for a remote supervisory computer. The ASCII format makes system debugging easy with a dumb terminal.

RS-232C is the most widely used communications standard for information transfer between computing equipment. RS-232C versions of the D1000 and D2000 will interface to virtually any computer without additional hardware. RS-232C is not designed to be used as a multiparty system; however the modules can be daisy-chained, as shown in figure 1, to allow many modules to be connected to a single communications port. In this network, any characters transmitted by the host are received by each module in the chain and passed on to the next station until the information is echoed back to the host. In this way all commands given by the host are examined by every module in the chain. If a module is correctly addressed and receives a valid command, it transmits a response on the daisy chain network. The response will be rippled through any other modules in the chain until it reaches the host.

RS-485 is a communications standard developed for multidropped systems that can communicate at high data rates over long distances, as shown in figure 2. RS-485 is similar to RS-422 in that it uses a balanced differential pair of wires switching from 0 to 5V to communicate data. RS-485 receivers can handle common mode voltages from -7

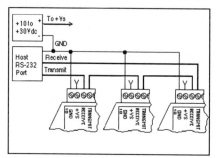

Figure 1 RS-232 Daisy Chain Network.

Figure C-1 (*Continued*)

DGH Corporation Modules

to +12V without loss of data, making them ideal for transmission over great distances. RS-485 differs from RS-422 by using one balanced pair of wires for both transmitting and receiving. Since an RS-485 system cannot transmit and receive at the same time it is a half-duplex system. For systems requiring many modules, high speed or long wiring distances the RS-485 standard is recommended.

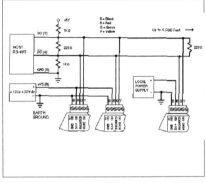

Figure 2. RS-485 Multidrop Network.

COMMAND SET

All DGH products use a simple command/response protocol for communication. A module must be interrogated by the host to obtain data. A module can never initiate a command sequence. A typical command/response sequence could look like this:

Command: $1RD
Response: *+00075.00

A command is initiated with a command prompt, which may be a dollar sign ($) or a pound sign (#). Following the prompt a single address character must be transmitted. Each module on a communications bus must be setup with a unique address. The command is directed in this case to module address '1'. The address is followed by a two-character command which in this case is RD for Read Data. The command is terminated with a carriage return.

After module address '1' receives the command it will respond with the analog input data. The response begins with a response prompt, which is an asterisk (*). The data is read back in a standardized format of sign, 5 digits, decimal point, and 2 more digits. All DGH modules represent data in the same standard format.

Table 1 shows all the D1000 and D2000 commands. For each case, a sample command and response is shown. Notice that some commands only respond with an * acknowledgment.

Figure C-1 *(Continued)*

Table 1. D1000 and D2000 Series Command Set.

Command and Definition		Typical Command Message ($ prompt)	Typical Response Message
DI	Read Alarms/Digital Inputs	$1DI	*0003
DO	Set Digital Outputs	$1DOFF	*
ND	New Data	$1ND	*+00072.00
RD	Read Data	$1RD	*+00072.00
RE	Read Event Counter	$1RE	*0000107
RL	Read Low Alarm Value	$1RL	*+00000.00 L
RH	Read High Alarm Value	$1RH	*+00510.00 L
RS	Read Setup	$1RS	*31070142
RZ	Read Zero	$1RZ	*+00000.00
WE	Write Enable	$1WE	*
Write Protected Commands.			
CA	Clear Alarms	$1CA	*
CE	Clear Events	$1CE	*
CZ	Clear Zero	$1CZ	*
DA	Disable Alarms	$1DA	*
EA	Enable Alarms	$1EA	*
EC	Events Clear	$1EC	*0000107
HI	Set High Alarm Limit	$1HI+12345.67L	*
LO	Set Low Alarm Limit	$1LO+12345.67L	*
RR	Remote Reset	$1RR	*
SU	Setup Module	$1SU31070142	*
SP	Set Setpoint	$1SP+00600.00	*
TS	Trim Span	$1TS+00600.00	*
TZ	Trim Zero	$1TZ+00000.00	*
D2000 Programming Commands (Write Protected).			
BP	Set Breakpoint	$1BP00-00200.00	*
EB	Erase Breakpoint Table	$1EB	*
MN	Set Minimum Value	$1MN-00200.00	*
MX	Set Maximum Value	$1MX+00750.00	*

For greater data security, options are available to echo transmitted commands and to send and receive checksums. The # command prompt requests a response message from the module that begins with an *, followed by the channel address, command, data (if necessary) and checksum. This response echoes the channel address and command for verification and adds checksum for error checking. Checksum is a two character hexadecimal value that can be added to the end of any command message, regardless of prompt, at your option. Checksum verifies that the message received is exactly the same as the message sent.

The DGH modules perform extensive error checking on commands and will respond with an error message if necessary. For example:

Command: $1AB
Response: ?1 COMMAND ERROR

All error messages start with an error prompt (?) followed by the channel address and error description. In this case, the module did not recognize 'AB' as a valid command.

162 Appendix C

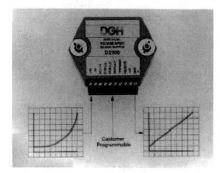

D2000 PROGRAMMING

The outstanding feature of the D2000 series is its user-programmable output scaling. The transfer function from analog input to data output may be specified to an infinite spectrum of functions, both linear and nonlinear. Sensor data may be scaled to any desired engineering units for easy interpretation.

The D2000 uses a piece-wise linear technique to approximate nonlinear functions. Figure 3 shows this technique. The first step in programming a function is to establish the functions endpoints, as shown in figure 3a. This is accomplished by using the Minimum (MN) and Maximum (MX) commands. In cases where only linear scaling is necessary, the programming task is now complete. For nonlinear functions, the linear curve may be broken into segments by describing a breakpoint using the BreakPoint (BP) command. The breakpoint establishes an intersection between two linear segments. Figures 3b & 3c show the effect of breakpoints.

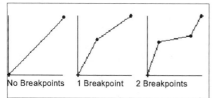

No Breakpoints   1 Breakpoint   2 Breakpoints

Figure 3. Piece-wise linear technique.

Up to 23 breakpoints are available to define 24 linear segments. Only two restrictions apply to the shape of the programmed transfer function:

1. The output data value must be a single-valued function of the input.
2. The output values must lie between the limits set by the endpoints.

Figure C-1 (*Continued*)

In general, breakpoints are defined by applying a known analog signal to the input of the module. This establishes the x-axis position of the breakpoint. The y-axis position is defined in the argument of the breakpoint (BP) command. The breakpoint data is stored in nonvolatile EEPROM. The transfer function may be reprogrammed many times.

RESOLUTION

All DGH modules represent data in the same fixed format of sign, five digits, decimal point, and two more digits; +00100.00 for example. The user can structure the D2000 output data for the best compromise between resolution and readability. For example, a +0.05 volt output indication may be structured in three output formats:

Input Voltage	Output Format	Resolution
+0.05 Volts	+00000.05	5
+50 millivolts	+00050.00	5,000
+50,000 microvolts	+50000.00	5,000,000

The microvolt output format extracts the best resolution but the output data will tend to be noisy. For a 0 to 0.05V output, millivolts is the best output format choice. This gives 5,000 counts of resolution in easy to interpret units.

In a typical application a D2000 module is used to output data in units of specific gravity. The specific gravity output range is between 0.5 and 2. If the output data format range is +0000.50 to +00002.00 there are only 150 counts of resolution between the minimum and maximum outputs. However, since the specific gravity of water is defined to be 1, the output may be scaled in percent. The specific gravity of water becomes 100 %. The output data range in % is from +00050.00 to +00200.00. This format allows up to 15,000 counts of resolution in easily interpreted units.

D2000 SCALING

The D2000 can output data in easy-to-understand engineering units that may be instantly read and interpreted, without data conversion, by a host computer. For example, a pressure sensor provides a 1 to 5V linear output for pressures of 0 to 1000 psi. A D2131 reads the sensor output in millivolts. But the real parameter of interest is pressure, not voltage, and voltage readings may be difficult to interpret. To make the output data more meaningful, program the D2131 output in psi:

Pressure (psi)	Sensor Output	D2131 Output (mV)	D2131 Output (psi)
0	1.0V	+01000.00	+00000.00
500	3.0V	+03000.00	+00500.00
1000	5.0V	+05000.00	+01000.00

In many cases, the desired output data is specific to an application. Assume that the same pressure sensor is used to measure the "fullness" of a pressure vessel, such as a cylinder of compressed air. The output units could be in units of "percent" and in this case we will assume that if the cylinder reads 750 psi it is 100% full:

Pressure (psi)	Sensor Output	D2131 Output (%)
0	1.0V	+00000.00
375	2.5V	+00050.00
750	4.0V	+00100.00

The real power of the D2000 is their ability to provide output data in engineering units for nonlinear sensors. A nonlinear transfer function may be programmed into a D2000 module by approximating the curve with a series of linear segments, using the Break Point (BP) command. A Break Point specifies the intersection between two linear segments. Up to 23 Break Points may be used to specify 24 linear segments in a curve.

The following example uses a D2131 to linearize the output of a pyrometer that uses an infrared temperature sensor. The infrared temperature sensor is inherently nonlinear and its output ranges from 0.717 to 1.406V for a temperature span of 600 to 1600°C.

Breakpoint	Input Voltage	Output Value
Minimum	+00717.00	+00600.00
00	+00844.00	+00700.00
01	+00948.00	+00800.00
02	+01036.00	+00900.00
03	+01110.00	+01000.00
04	+01174.00	+01100.00
05	+01230.00	+01200.00
06	+01280.00	+01300.00
07	+01325.00	+01400.00
08	+01367.00	+01500.00
Maximum	+01406.00	+01600.00

Scaling a nonlinear transfer function in the field
Assume that a water tower with an irregular shape is 30 feet tall and holds about 10,000 gallons. A pressure sensor may be used to measure the height of the water in the tower. The pressure sensor produces 0.1V per foot of water starting at 0V. To create a nonlinear function in the module, the endpoints must be set first. The minimum value is known and may be programmed by applying 0V to the module corresponding to 0 gallons. A "dummy" maximum value, which we know can never be exceeded, may be used to specify the maximum endpoint. In this case we apply +5V to the module and program the maximum value to be 15,000 gallons. Starting with an empty tower, read the pressure at fixed known volumes of water, every 1000 gallons for example, and set breakpoints in the module corresponding to known amounts of water in the tower. Once the curve is programmed, the module converts the pressure signal to gallons.

The preceding example shows that D2000 modules may be programmed in the field to specific test inputs where the actual nonlinearity is unknown. Since all programming

Figure C-1 (*Continued*)

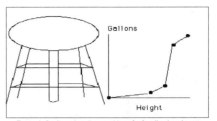

Figure 4. Scaling when the exact transfer function is unknown

is done through the communications port, access to a module is not necessary and ranging may be done remotely.

Scaling to desired engineering units
The D2000 allows you to scale an input to desired engineering units. For example, many sensor output signals are transmitted as 4 to 20mA signals. The following example demonstrates scaling a 4 to 20mA signal to 0 to 100% using a DGH D2251 or D2252 module. The actual input range of these modules is 0 to 25mA to make it easier to adjust for zero and span and to allow for drift in the end points of the input.

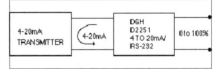

Figure 5. Scaling to desired engineering units

Since the input range is 0 to 25mA and you want to use a portion of that range, you must determine the new minimum and maximum values. The two desired values: 4mA, 0% and 20mA, 100% determines the desired transfer function. Extrapolate this function to the full-scale range of the module, which is 0-25mA. This results in endpoints at 0mA, -25% and 25mA, 131.25%.

Input the new minimum and maximum values with the following procedure. In these steps, we assume a channel address of 1.

1. Connect module to computer, or terminal and establish communications.
2. Apply 0mA to the input.
3. Send a Write Enable command, $1WE, followed by a Minimum Value command, $1MN-00025.00. The response to both commands should be an *.
4. Apply +25mA to the input.
5. Send a $1WE command followed by a Maximum Value command, $1MX+00131.25. The response to both commands should be an *.

The entire range is rescaled and all values are read in percent.

BP-8 and BP-14 8 and 14 CHANNEL MOUNTING BACKPLANES

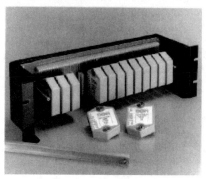

(Bracket not supplied)

The BP-8 and BP-14 are 8 and 14 channel mounting backplanes for DGH modules. The backplanes accept any RS-485 DGH analog input or analog output modules and are designed to be mounted in standard 19 inch racks. RS-485 modules are used because RS-485 is the preferred communications standard for high channel count applications. Although analog modules are used it must be noted that every DGH module has some digital I/O capability. Therefore the combination of DGH modules with the backplanes make a cost effective high density remote analog and digital data acquisition system.

The BP-8 and BP-14 reduce wiring costs by providing all common connections on the backplane. Each backplane includes screw terminals for all inputs, outputs, power connections and communications signals. The backplanes also include swaged thru-hole standoffs for mounting, a hold-down bar, and holes for an RS-485 termination resistor.

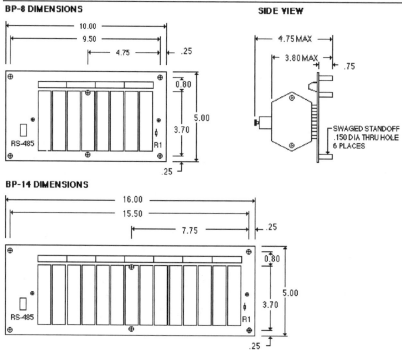

Figure C-1 (*Continued*)

ORDERING GUIDE

MODEL INPUT/OUTPUT
Voltage Input
D1101/D2101 10mV Input/RS-232C Output
D1102/D2102 10mV Input/RS-485 Output
D1111/D2111 100mV Input/RS-232C Output
D1112/D2112 100mV Input/RS-485 Output
D1121/D2121 1V Input/RS-232C Output
D1122/D2122 1V Input/RS-485 Output
D1131/D2131 5V Input/RS-232C Output
D1132/D2132 5V Input/RS-485 Output
D1141/D2141 10V Input/RS-232C Output
D1142/D2142 10V Input/RS-485 Output
D1151/D2151 100V Input/RS-232C Output
D1152/D2152 100V Input/RS-485 Output

Current Inputs
D1211/D2211 10mA Input/RS-232C Output
D1212/D2212 10mA Input/RS-485 Output
D1221/D2221 1mA Input/RS-232C Output
D1222/D2222 1mA Input/RS-485 Output
D1231/D2231 100mA Input/RS-232C Output
D1232/D2232 100mA Input/RS-485 Output
D1241/D2241 1A Input/RS-232C Output
D1242/D2242 1A Input/RS-485 Output
D1251/D2251 4-20mA Input/RS-232C Output
D1252/D2252 4-20mA Input/RS-485 Output

Thermocouple Inputs
D1311 J Thermocouple Input/RS-232C Output
D1312 J Thermocouple Input/RS-485 Output
D1321 K Thermocouple Input/RS-232C Output
D1322 K Thermocouple Input/RS-485 Output
D1331 T Thermocouple Input/RS-232C Output
D1332 T Thermocouple Input/RS-485 Output
D1341 E Thermocouple Input/RS-232C Output
D1342 E Thermocouple Input/RS-485 Output
D1351 R Thermocouple Input/RS-232C Output
D1352 R Thermocouple Input/RS-485 Output
D1361 S Thermocouple Input/RS-232C Output
D1362 S Thermocouple Input/RS-485 Output
D1371 B Thermocouple Input/RS-232C Output
D1372 B Thermocouple Input/RS-485 Output
D1381 C Thermocouple Input/RS-232C Output
D1382 C Thermocouple Input/RS-485 Output

MODEL INPUT/OUTPUT
RTD Inputs
D1411 .00385 RTD Input/RS-232C Output
D1412 .00385 RTD Input/RS-485 Output
D1421 .00392 RTD Input/RS-232C Output
D1422 .00392 RTD Input/RS-485 Output
D1431 .00388 RTD Input/RS-232C Output
D1432 .00388 RTD Input/RS-485 Output
D1451 2252Ω Thermistor Input/RS-232C Output
D1452 2252Ω Thermistor Input/RS-485 Output
D1461 TD Thermistor Input/RS-232C Output
D1462 TD Thermistor Input/RS-485 Output

Bridge Inputs
D1511/D2511 30mV Bridge Input, 5V Excitation/RS-232C Output
D1512/D2512 30mV Bridge Input, 5V Excitation/RS-485 Output
D1521/D2521 30mV Bridge Input, 10V Excitation/RS-232C Output
D1522/D2522 30mV Bridge Input, 10V Excitation/RS-485 Output
D1531/D2531 100mV Bridge Input, 5V Excitation/RS-232C Output
D1532/D2532 100mV Bridge Input, 5V Excitation/RS-485 Output
D1541/D2541 100mV Bridge Input, 10V Excitation/RS-232C Output
D1542/D2542 100mV Bridge Input, 10V Excitation/RS-485 Output
D1551/D2551 1-6V Bridge Input, 8V Excitation/RS-232C Output
D1552/D2552 1-6V Bridge Input, 8V Excitation/RS-485 Output
D1561/D2561 1-6V Bridge Input, 10V Excitation/RS-232C Output
D1562/D2562 1-6V Bridge Input, 10V Excitation/RS-485 Output

Timer and Frequency Inputs
D1601/D2601 Frequency Input/RS-232C Output
D1602/D2602 Frequency Input/RS-485 Output
D1611/D2611 Timer Input/RS-232C Output
D1612/D2612 Timer Input/RS-485 Output
D1621 Event Counter/RS-232C Output
D1622 Event Counter/RS-485 Output
D1631/D2631 Accumulator, Frequency Input/RS-232C Output
D1632/D2632 Accumulator, Frequency Input/RS-485 Output
D1641/D2641 Accumulator, Timer Input/RS-232C Output
D1642/D2642 Accumulator, Timer Input/RS-485 Output

Digital Inputs/Outputs
D1701 7 Digital Inputs, 8 Digital Outputs/RS-232C Output
D1702 7 Digital Inputs, 8 Digital Outputs/RS-485 Output
D1711 15 Digital Inputs and/or Outputs/RS-232C Output
D1712 15 Digital Inputs and/or Outputs/RS-485 Output

Figure C-1 (*Continued*)

Appendix D

Electrical Control Symbols

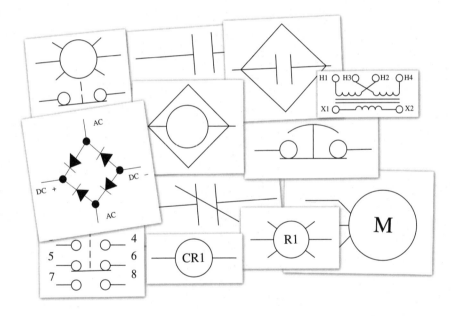

Symbols were developed to show the control logic on prints and diagrams. They are most used on ladder diagrams and wiring diagrams. Standards for symbols for electrical components have been established by the National Electrical Manufacturers Association (NEMA). Standards Publication ICS 19-2002 sets forth the standards for the use of symbols. A copy can be obtained at www.nema.org. Not all manufacturers, designers, and engineers use the standard symbols. Even when they use standard symbols, they sometimes use variations of the standard symbols. Of the 115 symbols that have been included here, some are not in the symbols standard but are commonly used on ladder diagrams.

Symbols are the language used on ladder diagrams and wiring diagrams. In spite of the fact that not all symbols are standard, knowledge of the symbols in this chapter will give you the ability to read electrical prints and ladder diagrams.

Designators

The following is a list of the standard designators used on electrical diagrams. Some of these can be used on an input or output.

Device or Function	**Designation**
Accelerating	A
Ammeter	AM
Braking	B
Capacitor	C or CAP
Circuit breaker	CB
Control relay	CR
Current transformer	CT
Demand meter	DM
Diode	D
Disconnect switch	DS DISC
Dynamic braking	DB
Field accelerating	FC
Field decelerating	FD
Field loss	FL
Forward	F or FWD
Frequency meter	FM
Fuse	FU
Ground protective	GP
Hoist	H
Jog	J

Electrical Control Symbols

Limit switch	LS
Lower	L
Main contactor	M
Master control relay	MCR
Master switch	MS
Overcurrent	OC
Overload	OL
Overvoltage	OV
Plugging or potentiometer	P
Power factor meter	PFM
Pressure switch	PS
Push button	PB
Reactor, reactance	X
Rectifier	REC
Resistor, resistance	R or RES
Reverse	REV
Rheostat	RH
Selector switch	SS
Silicon controlled rectifier	SCR
Solenoid value	SV
Squirrel cage	SC
Starting contactor	S
Suppressor	SU
Tachometer generator	TACH
Terminal block or board	TB
Transformer	T
Transistor	Q
Undervoltage	UV
Voltmeter	VM
Watthour meter	WHM
Wattmeter	WM

Figure D-1 Adjustability, general.

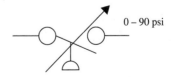

Figure D-1a Example: adjustability.

The symbol in Fig. D-1 is used to indicate that the device has an adjustable setting; the range is usually indicated on the diagram. In Fig. D-1*a* I am showing a pressure switch with a range of 0 to 90 psi.

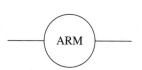

Figure D-2 Armature with commutator and brushes.

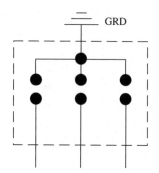

Figure D-3 Arrester, lightning three-pole.

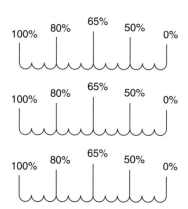

Figure D-4 Autotransformer.

Figure D-5 Battery. Polarity may or may not be shown, but the longer wire is always the positive lead.

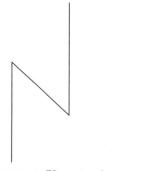

Figure D-6 Blowout coil.

Figure D-7 Brake coil.

Figure D-8 Capacitor.

In the capacitor symbol (Fig. D-8), the polarity may or may not be shown. The straight line is the +, or positive, lead. The curved line is the –, or negative, line. The polarity of the capacitor is relative to + and – leads. They can both be positive or negative relative to ground.

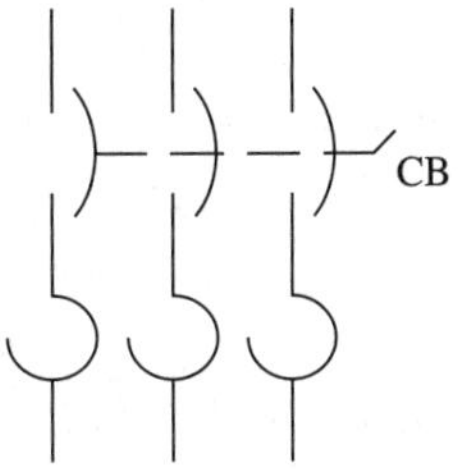

Figure D-9 Circuit breaker, three-pole with thermal trip units.

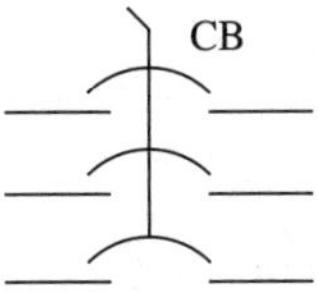

Figure D-10 Circuit breaker, three-pole, manual or nonautomatic.

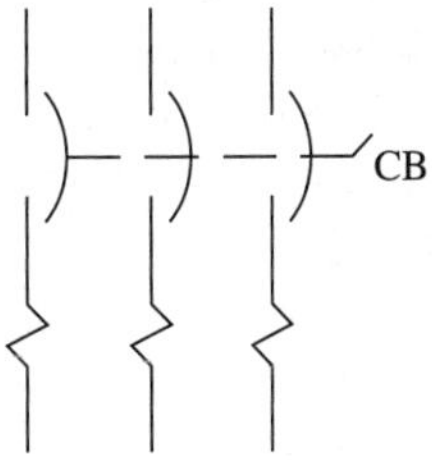

Figure D-12 Coil.

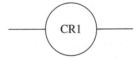

Figure D-11 Circuit breaker, three-pole, with magnetic trip units.

Figure D-12a Example: coil.

The symbol in Fig. D-12 is used to show a magnetic coil. It can be a relay, a motor starter coil, or a contactor. It can be any device that has a magnetic coil. In Fig. D-12a CR is for control relay and the number 1 is for number 1 control relay.

Typical coil and contactor designators

Function	Designation
Contactor	C
Latching	L
Main	M
Motor	Motor
Control relay	CR
Trip coil	TC
Unlatch coil	ULC

Figure D-13 Conductor.

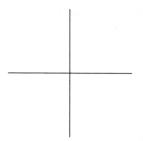

Figure D-14 Conductor, crossing not connected, not joined.

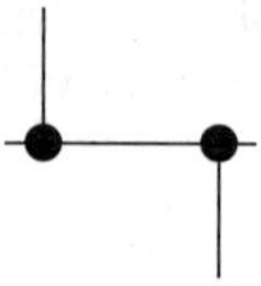

Figure D-15 Conductor, joined.

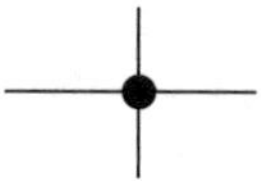

Figure D-15a Conductor, crossing joined.

Conductor, shielded

Shield may be shown grounded such as the one on the right. Only one side of a shield is grounded. This is to prevent a ground loop. Be careful to ground the correct end of the shield. If there are splices or junctions in the cable, make sure that the shield is spliced and does not contact ground. I like to use heat-shrink tubing to cover the shield at a junction or splice. Also keep the unshielded part of the conductor as short as possible to prevent noise on the conductor.

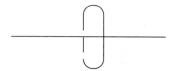

Figure D-16 Conductor, shielded.

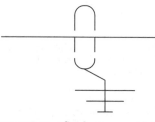

Figure D-16a Conductor, grounded shielded.

Figure D-17 Connection, mechanical.

Figure D-18 Connection, mechanical with interlock.

Figure D-19 Connector, female.

Figure D-20 Connector, male.

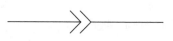

Figure D-21 Connector, separated or jacks enabled.

General information on contacts

Contacts are shown on a diagram or wiring print in their inactive state. This means that their operator is the deenergized or nonoperated position. The activating device may be of any type. It could be manually operated such as a push button or a foot switch; it could be mechanical such as a pressure switch. It could be electrically activated as a relay. A note might be provided to explain the proper point at which the contact is operated, for instance, the point where the contact opens or closes as a function of level, pressure, flow, voltage, current, etc. A note should be provided if the contact is shown in the energized state or where confusion may result. Actuating devices may have many contacts; any of them can be open or closed. For example, a relay may have an unlimited number of Normally Open or Normally Closed contacts that change state when the relay is energized. A pressure switch usually has a Normally Open contact and a Normally Closed contact. Care should be taken to select the correct contact.

The Normally Open and Normally Closed contacts shown below can be used to indicate any device. The designer can use the Normally Open contact to show a pressure switch contact. The designer will usually use a standard designator to indicate this, for example, PS next to the contact to indicate that it is a pressure switch. The number indicates that it is a contact of pressure switch 1. Typically there will be a note on the print giving the location of the pressure switch.

Figure D-22 Contact, example. PS1: Located on the left side of the machine on the incoming water supply line. The contact is Normally Open. Closed on pressure rise.

Figure D-23 Contact, Normally Closed.

Figure D-24 Contact, Normally Open.

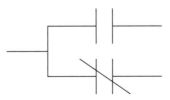

Figure D-25 Contact, transfer.

Time delay contacts

The direction of the arrow indicates the switch operation, in which the contact action is delayed (time delay closing and opening contacts).

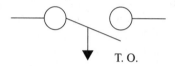

Figure D-26 Contact, Normally Open time delay closing (T.C.).

The contact in Fig. D-26 is open and will close after the timer is energized and the time that is set passes.

Figure D-27 Contact, Normally Open with time delay opening (T.O.).

The contact in Fig. D-27 is Normally Open. It will close when the timer is energized and will open after the time that is set passes.

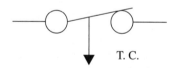

Figure D-28 Contact, Normally Closed with time delay opening (T.O.).

The contact in Fig. D-28 is Normally Closed and will open after the timer is energized and the time that is set passes.

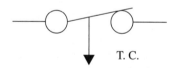

Figure D-29 Contact, Normally Closed with time delay opening (T.C.).

The contact in Fig. D-29 is Normally Closed. It will open when the timer is energized and will close after the time that is set passes.

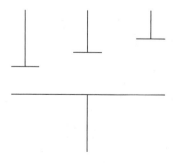

Figure D-30 Contacts, time sequential closing.

The contacts in Fig. D-30 are Normally Open and will close one at a time from left to right. Each contact will have a different time delay setting.

Figure D-31 Fuse.

Figure D-32 Generator, 3-phase.

Figure D-33 Ground, chassis or frame; a chassis, bus, or frame connection, which may be a substantial potential with respect to earth ground or structure on which it is mounted.

GRD

Figure D-34 Ground, chassis or frame; a chassis, bus, or frame connection that is at earth potenial.

Pilot light designators

These are used to show the color or type of the pilot light.

Designator	Color or Type
A	Amber
B	Blue
C	Clear
G	Green
NE	Neon
R	Red
W	White
Y	Yellow

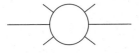

Figure D-35 Pilot light.

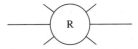

Figure D-35a Example: red pilot light.

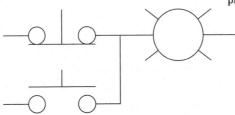

Figure D-36 Pilot light, push-to-test.

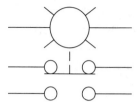

Figure D-37 Push-to-test pilot light with illuminated push button.

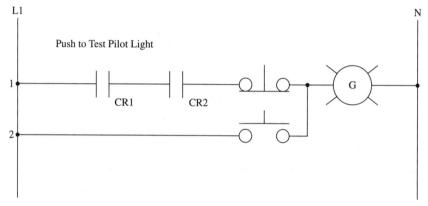

Figure D-38 Push-to-test pilot light, example.

Shown in Fig. D-38:

Line 1 The control relay contacts CR1 and CR2 must be closed, and the button not pushed. Under these conditions the green pilot light will light.

Line 2 If the button is pressed, the contact will close and the pilot will light.

Figure D-39 Meter.

To indicate the type of meter or instrument (see Fig. D-39), place the letter(s) from the list below in the symbol. For other functions, look for a note in the notes section.

Meter designators

AM	Ammeter	VA	Volt-ammeter
AH	Ampere-hour	VAS	Volt-ampere-reactive (VAR) meter
mA	Milliampere	VARH	VAR-hour meter
μA	Microampere	W	Wattmeter
PF	Power factor	WH	Watthour meter
V	Voltmeter		

Electrical Control Symbols

Mechanical motion

Figure D-40 Translation, one direction.

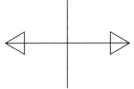

Figure D-41 Translation, both directions.

Figure D-42 Rotation, one direction.

Figure D-43 Rotation, both directions.

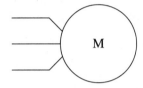

Figure D-44 Motor, 3-phase, induction.

Figure D-45 Magnetic overload relay: contact.

Figure D-45a Magnetic overlay relay: element.

Figure D-46 Thermal overload relay: contact.

Figure D-46a Thermal overload relay: element.

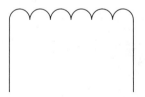

Figure D-47 Reactor (core not specified). Polarity may be added in a direct-current winding.

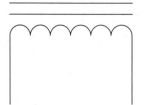

Figure D-48 Reactor, magnetic specified.

Rectifier, general

Arrow points in the direction of the current flow as indicated by a DC ammeter. Electron flow is in the opposite direction.

Figure D-49 Rectifier, semiconductor diode.

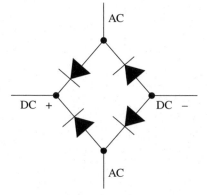

Figure D-50 Rectifier, full-wave-bridge-type.

Silicon-controlled rectifier (thyristor)

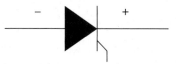

Figure D-51 P-type gate.

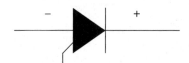

Figure D-51a N-type gate.

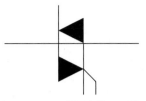

Figure D-52 TRIAC rectifier; bidirectional triode, thyristor.

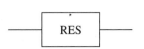

Figure D-53 Resistor, fixed.

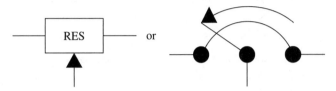

Figure D-54 Resistor, adjustable (rheostat).

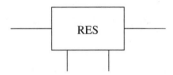

Figure D-55 Resistor, tapped.

Electrical or magnetic shielding

Figure D-56 Shielded.

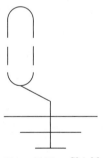

Figure D-56a Shielded with ground.

Figure D-57 Shunt, instrument.

A shunt (Fig. D-57) is typically used on an ammeter or a current transformer. A shunt should be installed on any secondary of a current transformer when it is not connected to a load.

Solid-state devices or elements

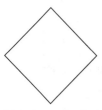

Figure D-58 Any logic or circuit symbol enclosed in a diamond indicates a solid-state device or element.

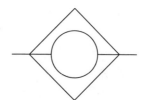

Figure D-59 Input.

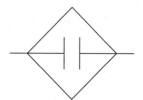

Figure D-59a Output.

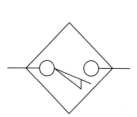

Figure D-60 Limit switch.

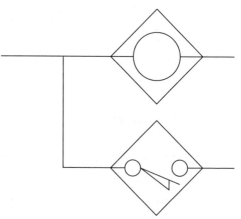

Figure D-61 When the output(s) and input(s) of solid-state devices are not isolated, the device will be enclosed in a box. The connections will be shown between the devices.

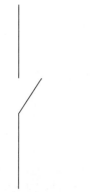

Figure D-62 Switch, single-throw.

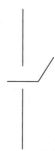

Figure D-63 Switch, double-throw.

Electrical Control Symbols

Figure D-64 Foot switch, opened by foot pressure.

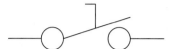

Figure D-65 Foot switch, closed by foot pressure.

Figure D-66 Limit switch, Normally Closed, held open.

Figure D-67 Limit switch, Normally Open, held closed.

Figure D-68 Limit switch, Normally Closed.

Figure D-69 Limit switch, Normally Open.

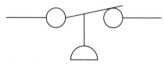

Figure D-70 Pressure switch, open by rising pressure.

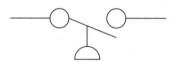

Figure D-71 Pressure switch, closed by rising pressure.

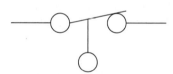

Figure D-72 Switch float, open on rising level.

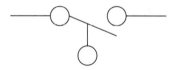

Figure D-73 Switch float, close on rising level.

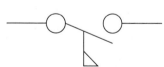

Figure D-74 Switch, flow close on flow (air, fluid, etc.).

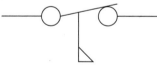

Figure D-75 Switch, flow open on flow (air, fluid, etc.).

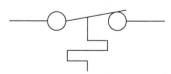

Figure D-76 Switch, temperature close on rising temperature.

Figure D-77 Switch, temperature open on rising temperature.

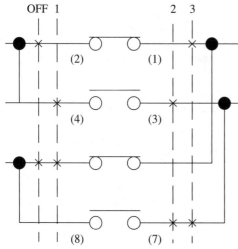

Figure D-78 Switch, master, four positions—Off, 1, 2, 3. X indicates contact closed in position.

Switch, push-button

Push buttons come in all shapes and sizes. There are only three types: Normally Open, Normally Closed, or a combination of contacts or illuminated. Most push buttons allow you to change the state of the contacts in the field. Many allow you to add contacts in the field, allowing for a large number of contacts either Open or Closed.

Figure D-79 Switch, push-button, Normally Open (momentary or spring-return).

Figure D-80 Switch, push-button, Normally Closed (momentary or spring-return).

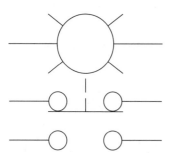

Figure D-81 Switch, push-button, illuminated, dual contacts (momentary or spring-return).

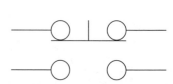

Figure D-82 Switch, push-button, dual contacts (momentary or spring-return).

Electrical Control Symbols 183

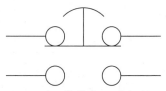

Figure D-83 Switch, push-button, dual contacts, with mushroom head (momentary or spring-return).

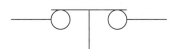

Figure D-84 Switch, wobble stick (momentary or spring-return).

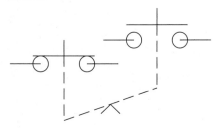

Figure D-85 Switch, push-button maintained, two circuit (latched or not spring-return).

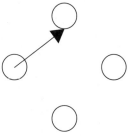

Figure D-86 Switches, selector nonshorting (nonbridging) during contact transfer.

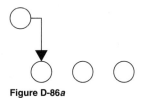

Figure D-86a

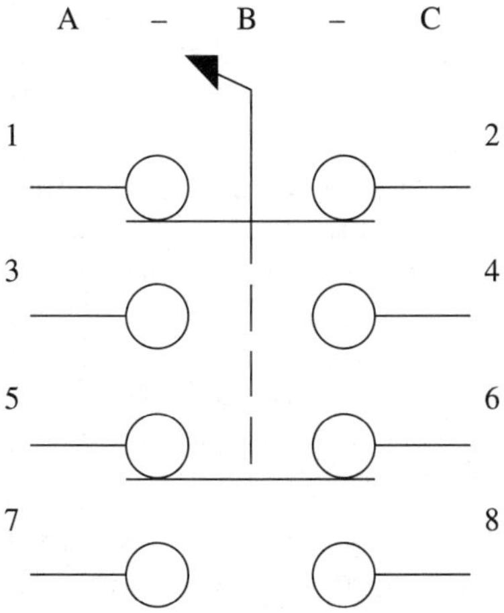

A – B – C

1　　　　　　　　2

3　　　　　　　　4

5　　　　　　　　6

7　　　　　　　　8

Figure D-87 Switch, selector push-button, with push-button type of contacts (can be pushed or pulled into two or more positions and rotated into two or more positions).

Appendix D

	Selector Position								
	A Button			B Button			C Button		
Contacts	In	Normal	Out	In	Normal	Out	In	Normal	Out
1–2	X			X	X		X		
3–4	X					X		X	X
5–6	X	X	X	X					
7–8							X	X	X

This table shows the state of each contact in Fig. D-87 in each position. A table like this will typically be in the Notes section of the drawing or as part of the specifications. If the selector or switch is programmed by the manufacturer, then it will be as an attachment to the specifications.

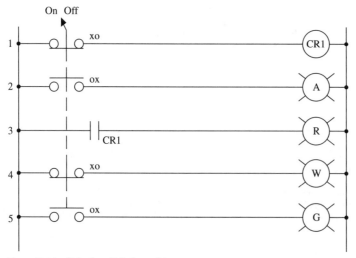

Figure D-88 Select switch 3-position.

The positions of the contacts can also be shown on a drawing like Fig. D-88 using Xs and Os. The X indicates contact closed and O indicates an open contact. The position of the X or O is the position of the switch, etc., A, B, C, or, if numbered, 1, 2, 3.

Line 1 If the switch is in position A, then according to the XOO next to the switch contact, the relay CR1 will be energized. If it is in position B or C, then the relay CR1 will be deenergized.

Line 2 If the switch is in position B, then according to the OXO next to the switch contact, the amber pilot light will be illuminated.

If it is in position A or C, then the amber pilot light will not be illuminated.

Line 3 If the relay CR1 is energized, then the Normally Open CR1 is closed and the red pilot light will be illuminated.

Line 4 If the switch is in position C, then according to the OOX next to the switch contact, the white pilot light will be illuminated. If it is in position A or B, then the white pilot light will not be illuminated.

Line 5 If the switch is in position B, then according to the OXO next to the switch contact, the green pilot light will be illuminated. If it is in position A or C, then the green pilot light will not be illuminated.

Speed switch
Operated by shaft rotation; F = forward, R = reverse.

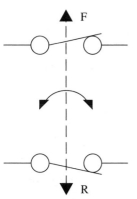

Figure D-89 Antiplugging switch (prevents plugging of drive).

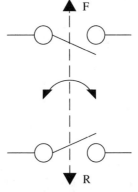

Figure D-89a Plugging switch (remove plug-stop action after drive has practically come to rest).

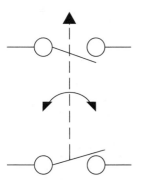

Figure D-90 Preset speed switch (operates at a preset speed).

Figure D-91 Synchronous motor or AC generator. Omit the field for synchronous induction, reluctance, or hyteresis motor.

Figure D-92 Terminal. Indicates that there is a field-accessible terminal; may be on a terminal point on a device or terminal strip.

Figure D-93 Terminal. Indicates that there is a field-accessible terminal on a terminal strip.

Figure D-93a Example of terminal in Fig. 93.

Figure D-94 Terminal, wire connection. Indicates a connection of the wires that is not field terminal.

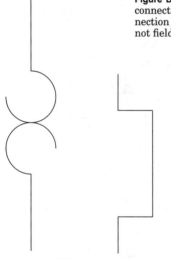

Figure D-95 Thermal.

Figure D-95a Element actuating device.

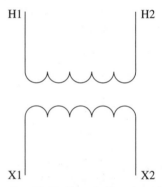

Figure D-96 Transformer, general (core not specified). On elementary diagrams, windings of the same transformer may be shown at different locations.

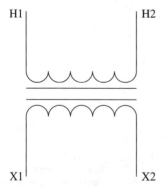

Figure D-97 Transformer, magnetic core.

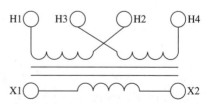

Figure D-98 Transformer, multitapped primary (magnetic core).

In Fig. D-98, connect H3 to H2 for 220 to 110 V. For an isolation transformer connect H3 to H1 and H2 to H4 for 110-V input to 110-V output. Always check the nameplate on the transformer to make sure that is the correct voltage.

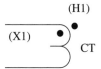

Figure D-99 Transformer, current.

In Fig. D-99, the instantaneous direction of the current into one polarity mark (•) corresponds to the current out of the other polarity mark.

Care should be taken with the secondary of the current transformer. The secondary should not be disconnected from the load while current is flowing. The transformer will attempt to continue driving current across the effectively infinite impedance. This will produce a high voltage across the secondary, which can be in the range of several kilovolts and can cause arcing. This voltage will compromise your safety and permanently affect the accuracy of the transformer.

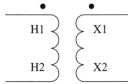

Figure D-100 Transformer, potential.

In Fig. D-100, the instantaneous direction of the current into one polarity mark (•) corresponds to the current out of the other polarity mark.

Figure D-101 Winding.

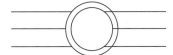

Figure D-102 Wound rotor induction motor or induction frequency converter.

Index

A3000, 128
AC. *See* alternating current
accumulator input, 158
Alarm Active pilot light, 13, 20
alarm outputs, 160
Allen Bradley 837E temperature sensor, 111, 119, 123
Allen Bradley Micrologix 1100, 104–105, 111, 119, 123
    RS-232 conversion and, 127–128
    RS-485 conversion and, 127–128
alternating current (AC), 2, 8, 131
    control circuit, 23
amber pilot light, 13, 69
ammeter, 131
ampere, 131
anode, 131
antiplugging switch, 185
armature, 170
armed mode, 19
    EPO, 6
arrester, 170
ASCII, 128
asynchronous, 107
Auto mode, 21, 29
auto position, 4-position Selector switch devices in, 47, 54–55
automatic float switch, 60–62
autotransformer, 170

battery, 170
BCD. *See* Binary Coded Decimal
bill of materials, 18
Binary Coded Decimal (BCD), 117
black power negative, 116
blower circuits, 68
blowout coil, 170

bolts
    dead, A, 56
    dead, B, 56
    spring, 56
booted unit, 36
bootless flush head unit, 36
BP-8, 164
BP-14, 164
brake coil, 170
branch circuit, 131
bridge input, 158, 165
brushes, 170
bulletin 836T pressure controls, 63
bulletin 837, 66
buttons. *See specific buttons*

calorie, 131
capacitance, 131
capacitative proximity sensors, 74–77
capacitative reactance, 132
capacitor, 132, 170
CB, 143
chassis, 175
circuit (series), 132. *See also specific circuits*
circuit breakers, 132, 143
    three-pole, 171
closed circuit, 132
cluster pilot light devices, 53
coil, 132
    blowout, 170
    break, 170
command set, 161
commutator, 170
computer room air conditioning (CRAC), 17, 105
    Modbus protocol and, 126
    shunt trip circuits, 11

conductance, 132
conductors, 132
  joined, 172
  shielded, 172
conduit entrance connection, 63
contact terminal 3, Normally Open, 83
contact terminal 4, Normally Open, 83
contact terminal 5, Normally Closed, 83
contact terminal 6, Normally Closed, 83
contacts, 173
control voltage, 2
conveyor belts, 72
conveyor circuit, 70, 76
convolution bellows, 63
Coulomb, 133
counter EMF, 133
CPC shunt trip circuits, 11
CR1, 35, 37, 39, 40, 73, 75, 79, 82, 176, 184–185
  Normally Open contact of, 85, 86–87, 96, 112, 124
  relay, 82, 83
CR2, 79, 100, 176
CRAC. *See* computer room air conditioning
crimp-type wire nut, 109–110
Critical Power Lost pilot light, 14, 20
cross-circuits, 133
current, 133
  input, 157, 165
  transformers, 187
  unit of, 140
customer marking area, 63
cycle, 133
cylinder lock operator, 41, 45
  mushroom style, 56
  push-button, 56
  3-position, 43

D1000, 157–165
  alarm outputs, 160
  command set, 161
  communications, 160–161
  digital filter, 160
  digital inputs/outputs, 159
  event counter, 160
  general description, 159
  resolution, 162
  theory of operation, 159
  user options, 160
  utility software, 159
  voltage inputs, 157
D1100, 157
D1200, 157
D1300, 157
D1400, 157, 160

D1450, 158
D1500, 158, 160
D1600, 158
D1621, 160
D1622, 160
D1700, 158
D1701, 158
D1702, 158
D2000, 157–165
  alarm outputs, 160
  command set, 161
  communications, 160–161
  digital filter, 160
  digital inputs/outputs, 159
  event counter, 160
  general description, 159
  programmable features, 156
  programming, 162
  resolution, 162
  scaling, 162–163
  theory of operation, 159
  user options, 160
  utility software, 159
D2100, 157
D2200, 157
D2500, 158, 160
D2600, 158
data communication equipment (DCE), 106
data terminal equipment (DTE), 106, 114
dB. *See* decibel
DC. *See* direct current
DCE. *See* data communication equipment
dead bolt A, 56
dead bolt B, 56
deci, 133
decibel (dB), 133
deflection, 133
deka, 133
designators, 168
  meter, 176–177
  pilot light, 175–176
DGH Corporation, 105, 108, 155–165
DGH1, 125
  analog input to, 112, 120, 124
  power for, 112, 120, 124
diagram, 133
digital filter, 160
digital inputs, 158, 160, 165
digital outputs, 158, 160
diode, 133
direct current (DC), 2, 134
dissipation, 134
double-pole double-throw (DPDT), 84, 87, 90
  with dual coils, 90

double-pole single-throw with one Normally Open contact (DPST-NO), 84
double-throw switches, 180
DPDT. *See* double-pole double-throw
DPST-NO. *See* double-pole single-throw with one Normally Open contact
drop, 134
DTE. *See* data terminal equipment

earth, 134
eddy current, 71–72
EEPROM, 159, 160
800T push button, 32
837E temperature sensor, 111, 119, 123
8-pole relay, 8
electric circuit, 134
electrical control symbols, 167–187
electrical horsepower, 134
electrical shielding, 179
electrical units, 134
electrician, 134
electricity, 134
electrocution, 134
electromagnetic, 71, 134
electromotive force (EMF), 135
    counter, 133
emergency power off (EPO), 99, 145–150
    armed mode, 6
    maint mode, 7
    Modbus armed input, 15–16
    Modbus main mode input, 16
    Modbus power on input, 16
    Modbus Rest input, 15
    Modbus test input, 15
    procedure, 19–20
    system ladder diagram, 3–18
    test mode, 7
emergency stop, 56
EMF. *See* electromotive force
energy efficiency, 135
EPO. *See* emergency power off
Ethernet TCP/IP, 122
    RS-232 and, 127
    RS-485 and, 127
    specifications, 127–128
event counter, 160
    input, 158
extended head unit, 36
    without guard, 37

Faraday effect, 135
fathom, 135
fiber optics, 135
fire alarm, 99
flip lever operator, 58

float operator assembly, 62
float switch, 44, 54, 89
    automatic, 60–62
fluorescence, 135
flush head unit, 36
force, 135
formula, 135
4-position Selector switch devices, 41, 44–45
    in Auto position, 47, 54–55
    in Hand position, 46, 54
4-position toggle switch, 53
frequency, 135
    input, 158, 165
front panel display, 18, 30
FS1, 85, 89
    Normally Open, 82
FU, 101, 143
fuse, 136

galvanometer, 136
green pilot light, 69, 82
green signal, 116
ground terminal, 60
grounded, 136

H1, 101, 187
H2, 187
H3, 187
H4, 101, 187
Hand mode, 21, 23, 24
Hand position, 42
    4-position Selector switch devices in, 46, 54
    3-position Selector switch in, 42–43
Hand/Off/Auto Selector switch, 61–62, 64–65, 67–68, 70, 73–74, 75
heat, 136
heat shrink tubing, 111
heavy-duty industrial relays, 85–87
horsepower, 136
    electrical, 134

IBM-AT, 108
ice cube, 87–90
illuminated 2-position push-pull, 36
illuminated 2-position push-pull/twist, 36
illuminated 2-position Selector switch operators, 47
illuminated 3-position push-pull, 50
illuminated 3-position Selector switch, 48
impedance, 136
inductance, 136
induction, 137
inductive proximity sensors, 71–74

inputs, 137, 180
  accumulator, 158
  bridge, 158, 165
  current, 157, 165
  devices, 31–80
  digital, 158, 160, 165
  event counter, 158
  frequency, 158, 165
  RTD, 157, 165
  thermistor, 158
  thermocouple, 157, 165
  timer, 158, 165
  voltage, 157, 165
instantaneous contacts, 92
insulator, 137
interface communication, 107
ion, 137
IPS1, 71, 75
IT Equipment Room, 4, 20, 35, 99

joint, 137
Joule's law, 137

keyed Selector switch, 8, 9, 19, 40, 41
  3-position, 48
kilovolt (kV), 137
kilowatt (kW), 137
  meter, 126
knob lever operator, 39, 42, 44, 47
kV. *See* kilovolt
kW. *See* kilowatt

ladder diagrams, 1–30, 145–153
  EPO, 3–18
  punch press, 20–30
  simple, 2, 3
LAN. *See* local-area network
latch push button
  Normally Closed, 91
  Normally Open, 91
latching relays, 90–91
leakage, 137
LED, 68
limit switches, 25–27, 29, 77–80, 180–181
  quick selection guide to, 77
line noise, 107
local-area network (LAN), 127
locking nut, 63
logic reed block, 33, 34
loss, 137
low-voltage control circuit, 99
LS1, 79
LS2, 79–80
LS3, 79
LS4, 79
LS5, 79
LS6, 79

M1, 3, 38, 48, 54, 55–56, 82, 93, 143
  Normally Closed overload contact of, 96
  Normally Open contact of, 52, 57–58, 74, 102
M2, 46, 54
magnetic core transformers, 186
magnetic overload relay, 177
magnetic shielding, 179
maint mode, 14, 20
  EPO, 7
  Modbus EPO, input, 16
Maintenance Mode pilot light, 14
mechanically interlocked maintained push-button devices, 52
megavolt, 137
meter, 137
  designators, 176–177
  galvanometer, 136
  kilowatt, 126
  potentiometers, 51
Micrologix 1100, 104–105, 111, 119, 123
  RS-232 conversion and, 127–128
  RS-485 conversion and, 127–128
mini contact block, 34
Modbus, 20, 128
  altering, 12
  CRAC units and, 126
  EPO armed input, 15–16
  EPO maint mode input, 16
  EPO power on input, 16
  EPO reset input, 15
  EPO test input, 15
  Normally Open output from, 9
  output contacts, 125–126
  power for, 14
  specifications, 122–126
module address, 117
momentary contact push-button devices, 39
momentary wobble stick push-button device, 58–59
monitoring systems, 103–129
  control modules, 105–106
  defined, 104
motor belt covers, 29
motor control, three phase, 143
motor covers, 29
motor starters, 28
  circuits, 101
motor thermal overloads, 143
mounting bracket, 60
multitapped primary transformers, 186
mushroom-style cylinder lock, 56
mutual induction, 140

Index 193

nameplate, 66
National Electrical Manufacturers Association (NEMA), 2, 168
NEC 430-72, 7, 23
NEC 645, 4, 8
NEC 2008 Article 230.3 D-3, 23
NEC Section 645.10, 99
negative, 137
NEMA. *See* National Electrical Manufacturers Association
network, 138
neutron, 138
nonlinear transfer function, 163
Normally Closed, 2, 8, 41
  contact of relay CR1, 83
  contact R9, 28
  contact terminal 5, 83
  contact terminal 6, 83
  contacts, 173–174
  latch push button, 91
  Mushroom push button, 100
  overload contact of M1, 96
  push buttons, 55, 143
  Stop button, 51, 57, 61–62, 86, 102
  Stop Red Mushroom push button, 24, 27
Normally Open, 3, 8, 27, 41
  contact of CR1, 85, 86–87, 96, 112, 124
  contact of M1, 52, 57–58, 74, 102
  contact of R1, 10, 12
  contact of R2, 10
  contact of R4, 11
  contact of relay R3, 15
  contact of relay R5, 13
  contact of relay R6, 11
  contact of relay R7, 11, 13, 15–16
  contact of relay R8, 11
  contact of relay R9, 14
  contact of relay R10, 16
  contact of SSR1, 96, 112
  contact terminal 3, 83
  contact terminal 4, 83
  contact terminal 5, 125
  contact terminal 6, 125
  contact terminal 7, 125
  contact terminal 8, 125
  contacts, 173–174
  contacts of Selector switch, 12
  contacts of wobble switch, 59
  flow FS1, 82
  Green Start button, 24
  latch push button, 91
  Mushroom keyed release push buttons, 8
  Mushroom push button, 100
  output from Modbus, 9
  push buttons, 143

Normally Open (*Cont.*):
  single-pole single-throw double-make with one, contact, 83, 87–88, 95
  Start button, 52, 57–58, 73
  time-close contact, 93
  time-close contact of SSTR1, 99
N-type gate, 178

Off mode, 21, 23, 24
ohm, 138
OL, 143
100 BASET 100 Mbit/s, 127
OOOX, 44, 46, 54
OOX, 12, 49, 70, 76, 78, 185
OOXO, 44
open circuit, 138
open frame relay, 83–85
operating lever, 60
operating range adjustment screw, 63, 66
operation indicator, 63
orange signal, 116
ordering guide, 165
out position, 3-position push-pull in, 49–50
output(s), 138, 180
  alarm, 160
  devices, 82
  digital, 158, 160
output contacts, 83
  Modbus, 125–126
  RS-232, 114
  RS-485, 121–122
over-center toggle mechanism, 60
OXO, 10, 69, 75, 79, 184, 185
OXOO, 44, 46, 54

padlocking jumbo mushroom button, 57
padlocking mushroom button, 57
panel terminal strip, 113
peak, 138
  current, 138
pen tuff (low-voltage) block, 33, 34
period, 138
photoelectric sensor, 68–71, 138
photosensors, 68
piece-wise linear technique, 162
pilot light, 12, 24, 26, 29, 38
  Alarm Active, 13, 20
  amber, 13, 69
  cluster, 53
  Critical Power Lost, 14, 20
  designators, 175–176
  green, 69, 82
  Maintenance Mode, 14
  optional, 63
  push-to-test, 175–176
  Test complete, 13

194 Index

PLC. *See* programmable logic controller
plugging switch, 185
pneumatic timers, 92
positive, 139
potential transformers, 187
potentiometers, 51, 97
   time cycles set by, 52
power, 139
Power On, 7, 19
   indicator, 23
preset speed switch, 185
pressure, unit of, 140
pressure switch, 62–65, 181
   bulletin 836T, 63
programmable logic controller (PLC), 104–105
proton, 139
proximity sensors, inductive, 71–74
P-type gate, 178
punch press, 20–30
   procedure, 28–30
   system, 151–153
   three-sheet, 21
push buttons, 32–33
   cylinder lock, 56
   dimensions, 34
   800T, 32
   latch, 91
   momentary wobble stick, 58–59
   Normally Closed, 55, 143
   Normally Closed Mushroom, 100
   Normally Open, 143
   Normally Open Mushroom, 100
   Shutdown, 100
   switches, 182–185
   2-position Selector switches with, 55

quick-break, 139

R1, Normally Open contact of, 10, 12
R2, 9
   Normally Open contact of, 10
R3, 9
R7, 17
R8, closure of, 17
R9
   closure of, 17
   Normally Closed contact, 28
range scale, 63
reactance, 139
rectifiers, 178–179
   silicon-controlled, 178–179
red power positive, 116
reed relay, 94

relay(s), 82, 139. *See also specific relays*
   8-pole, 8
   heavy-duty industrial, 85–87
   latching, 90–91
   magnetic overload, 177
   open frame, 83–85
   reed, 94
   sealed, 87–90
   solid-state, 68, 69, 74, 75, 94–97
   solid-state timer, 97–99
   thermal overload, 177
   time delay, 75
   timing, 92–94
relay CR1
   Normally Closed, 83
   Normally Open contact of, 82
relay R2, 24
relay R3, Normally Open contact of, 15
relay R5, Normally Open contact of, 13
relay R6, 10
   Normally Open contact of, 11
relay R7, Normally Open contact of, 11, 13, 15–16
relay R8, Normally Open contact of, 11
relay R9, 12, 27
   Normally Open contact of, 14
relay R10, 12
remote terminal unit (RTU), 114
reset push button, 9, 19, 20, 37, 38, 39
resistance, 139
   unit of, 140
resistor, 139, 179
resolution, 162
RightSight DC model, 69
Rockwell Automation's Allen Bradley devices, 32, 82
rotation, 177
RS-232, 104, 159, 160
   conversion from, 127–128
   daisy-chain network, 108
   DB9 pinout, 107, 108
   DB25 pinout, 107, 108
   Ethernet and, 127
   loop network, 110–111
   original pin-out for, 108
   output contacts, 114
   shielded twisted-pair cable for, 110–111
   specifications, 106–113
   wiring diagram, 109
RS-485, 104, 105, 126, 159
   conversion from, 127–128, 129
   Ethernet and, 127
   module address, 117
   output contacts, 121–122

Index   195

RS-485 (*Cont.*):
  shielded twisted-pair cable for, 118–119
  specifications, 114–122
  wiring diagram, 116–117
RS-3232, 105
RTD input, 157, 165
RTU. *See* remote terminal unit

scaling, 162–163
Schneider Electric, 122
SCSI. *See* small computer system interface
sealed relay, 87–90
sealed switch block, 33, 34
  stackable, 34
select switch, 184
Selector switch, 10, 32–33
  4-position, 41
  Hand/Off/Auto, 61–62, 64–65, 67–68, 70, 73–74, 75
  illuminated 3-position, 48
  keyed, 8, 9, 19, 40, 41, 48
  Normally Open contacts of, 12
  3-position, 24, 42–43
  2-position, 47
semiconductor, 139
serial, 107
series circuit, 139
series parallel circuit, 139
700-HG, 84
700-HS, 94
700-HV, 94
700-SE, 112, 119, 123
shallow block, 33, 34
shielded cables, 109
shielded twisted-pair cable
  for RS-232, 110–111
  for RS-485, 118–119
  terminating, 110–111, 118–119
short circuit, 139
shunt, 140
shunt trip breakers, 4, 99–100, 140
shunt trip circuits
  CPC, 11
  CRAC, 11
Shutdown push button, 100
silicon-controlled rectifier, 178–179
simple ladder diagrams, 2, 3
single-pole double-throw (SPDT), 84, 90, 92, 95
single-pole single-throw double-make with one Normally Open contact (SPST-NO-DM), 83, 87–88, 95
single-throw switches, 180

small computer system interface (SCSI), 129
snap action contact block, 34
solenoid, 140
solid-state coils, 97
solid-state devices, 180–187
solid-state relays, 68, 69, 74, 75, 94–97
  generating heat, 95
solid-state timer relays, 97–99
  power to, 98–99
solid-state timers, 92
SPDT. *See* single-pole double-throw
speed switch, 185–187
spring bolt, 56
SPST-DO-DM. *See* single-pole single-throw double-make with one Normally Open contact
SS-Cont, 10
SSR1, 96, 124
  Normally Open contact of, 96, 112
  power for, 112, 120
SSTR1, 98–99
  Normally Open time-close contact of, 99
  temperature switch triggering, 98
stackable sealed switch block, 34
standard knob operator, 39, 42, 44, 47, 48
standard symbols, 2
Start button, 3, 37, 38, 54
  Normally Open, 52, 57–58, 73
Start-Stop station, 52
steady current, 140
Stop button, 3, 37, 38, 42
  Normally Closed, 51, 57, 61–62, 86, 102
straight pair cable, 115
sump operation, 61, 62
switches. *See specific types*
synchronous motor, 185

tandem mounting, 34
tank operation, 61, 62
temperature sensor, 120
temperature switch, 65–68
  triggering SSTR1, 98
terminal 54, 11
terminal 55, 11
terminal strip, 25
Test Complete pilot light, 13
test mode, 19
  EPO, 7
thermal overload relay, 177
thermistor input, 158
thermocouple input, 157, 165
3-Cont, 10, 13, 14
3PDT. *See* three-pole double-pole double-throw

3-phase motor control, 143
3-position cylinder lock operator, 43
3-position keyed Selector, 48
3-position metal push-pull, 35
3-position push-pull, 49
   illuminated, 50
   in Out position, 49–50
3-position Selector switch, 42–43
   in Hand position, 42–43
   illuminated, 48
three-phase motor starter, 100–102
three-pole circuit breakers, 171
three-pole double-pole double-throw (3PDT), 87
three-sheet punch press, 21
three-wire motor control circuit, 3
three-wire start/stop station, 86
time delay contact block, 34
time delay contacts, 174–175
time delay relay, 75
time-close contact, Normally Open, 93
timer inputs, 158, 165
timing relays, 92–94
toggle switch, 4-position, 53
TR1, 51, 75, 77, 79–80, 93
transformers, 140
   current, 187
   magnetic core, 186
   multitapped primary, 186
   potential, 187
transistors, 94, 140
translation, 177
TRIAC, 94, 179
twisted-pair cable, 116
twisted-pair shielded cables, 107, 116
2-Cont, 10, 13, 14
2-position push-pull, 35
   illuminated, 36

2-position push-pull/twist, 35
   illuminated, 36
2-position Selector switch operators, 47
   with push-button operators, 55

unit of current, 140
unit of pressure, 140
unit of resistance, 140
universal serial bus (USB), 129
   conversion to, 129
USB. *See* universal serial bus
user options, 160
UTP CAT-5, 106

volt (V), 140
voltage difference, 115
voltage drop, 141
voltage input, 157, 165
voltage level, 115

watt (W), 141
winding, 187
wiring diagrams, 72
   of RS-232 daisy-chain network, 109
wobble stick switch, 183
wobble stick unit, 58–59
   Normally Open contacts of, 59
wound rotor induction motor, 187

X1 terminal, 101
X2 terminal, 101
XO, 55
XOO, 24, 42, 43, 48, 49, 50, 61, 62, 69, 74, 75, 79, 184
XOOO, 44, 46, 54

Y connection, 141